Methods in Molecular Biology

Series Editor
John M. Walker
School of Life and Medical Sciences
University of Hertfordshire
Hatfield, Hertfordshire, AL10 9AB, UK

For further volumes:
http://www.springer.com/series/7651

Plant Receptor Kinases

Methods and Protocols

Edited by

Reidunn Birgitta Aalen

Department of Biosciences, University of Oslo, Oslo, Norway

Humana Press

Editor
Reidunn Birgitta Aalen
Department of Biosciences
University of Oslo
Oslo, Norway

ISSN 1064-3745 ISSN 1940-6029 (electronic)
Methods in Molecular Biology
ISBN 978-1-4939-7062-9 ISBN 978-1-4939-7063-6 (eBook)
DOI 10.1007/978-1-4939-7063-6

Library of Congress Control Number: 2017942947

Printed on acid-free paper

This Humana Press imprint is published by Springer Nature
The registered company is Springer Science+Business Media LLC
The registered company address is: 233 Spring Street, New York, NY 10013, U.S.A.

Preface

Investigating Plant Receptor Kinases

During development, each plant cell needs to know its position in the organ where it resides to express the appropriate genes that will give the cell the ability to perform its specific function. Over the last decade, it has become increasingly more evident that—in addition to the classical hormones—the necessary positional information is provided by the activation of plasma membrane-bound receptor-like kinases. Furthermore, plant receptor kinases (PRKs) are involved in the sensing and response to environmental cues. Thus, plant scientists have revealed that signaling via receptor kinases is crucial both in developmental processes and in response to biotic and abiotic stress. Still, there are substantial challenges: hundreds of PRKs are without known ligand and without known function; furthermore, co-receptor identity and interaction mode, as well as signal transduction mechanisms, need to be elucidated and understood.

This volume of *Methods in Molecular Biology* presents protocols relevant for PRK both in development and defense and covers each step in a comprehensive experimental approach for molecular investigations of plant receptor kinases, from the simplest methods for expression and purification of receptor domains to the most advanced methods aiming at understanding the dynamics of receptor complex formation and specificity of signaling pathways.

The methods have been divided into sections, starting with the different protocols for expression and purification of full-length receptors, or specific domains. The PRKs are most often expressed in specific tissues or cells at low concentrations and are therefore difficult to purify from the plant in sufficient amounts for molecular and biochemical experiments and investigations. To study PRKs, it is therefore usually necessary to express the protein in a heterologous system.

E. coli can preferentially be used to express and purify PRK domains, especially the kinase domain, rather than the full-length receptors, which often fail to express at all (Chapter 1). Advice is given for cloning domains, vector choice, expression conditions, and handling of inclusion bodies. When attempting to express a domain, correct delineation of the domain is crucial, and structural information can be of good use. One has to be aware that a protein expressed in *E. coli* will not get the posttranslational modifications found in planta.

Using a simple eukaryote like yeast may solve such problems. PRKs retain their structure and find their way to the plasma membrane when expressed in yeast, which is a simple and cheap expression system (Chapter 2). A vector system is suggested, and the PRK can either be studied in the yeast plasma membrane or be isolated from the microsomal fraction for biochemical experimentation.

A cell culture alternative to yeast is the use of tobacco BY-2 cells (Chapter 3). The BY-2 cells grow slower than yeast but have the advantage of being transformable with *Agrobacterium tumefaciens*, using vector systems familiar to plant science labs. Since the BY-2 cells as plant cells can be expected to fold the expressed receptor correctly, they are especially attractive for testing of binding of potential ligands to the ectodomain. For that purpose, it is advisable to express a receptor with a deleted kinase domain, as this has been shown to give a higher and more stable expression.

For some purposes, like structural analysis or proteomics, the methods presented in Chapters 1–3 may not be producing sufficient material. Therefore, a detailed protocol for a wheat germ cell-free expression system is included (Chapter 4). A number of improvements over 40 years' time make it now possible with continuous production over 60 h which can provide more than 1 mg of protein. The bilayer cell-free wheat germ expression method has eliminated the need of dialysis, and the "split primer" method provides a reliable way of making constructs for efficient transcription. In the cell-free system, it is furthermore easy to modify the reaction conditions or introduce modifications in the protein.

PRKs are involved in cell-to-cell communication in multicellular organism, so whether efficient protein expression is in the cell-free system or in cell culture from heterologous hosts, the ultimate achievement would be to isolate the receptor from its natural host under natural conditions. This is a challenge that Chapter 5 discusses, along with two methods: the extraction of protein from isolated microsomal membrane fractions of *Arabidopsis* or *Nicotiana benthamiana* leaves expressing a tagged receptor, followed by immunoprecipitation, and alternatively the batch method with purified receptor kinase immobilized on a biosensor chip.

Such a chip can be used for investigating binding affinities and kinetics of the ligand-receptor pairs. Ligand binding and receptor activation is addressed in Part II. Candidate ligands of a given receptor are often identified based on genetic evidence. These putative signaling partners must be proven to bind and activate the receptor. This usually requires a label (e.g., radioactive ^{125}I or acridinium-ester) on the potential ligand, conditions that stabilized the ligand-receptor interaction, and a reliable readout. Chapter 6 provides a detailed protocol for using photoaffinity labeling to achieve covalent binding between ligand and receptor and thereby compelling evidence for their interaction. Other methods are needed to show activation of the receptor. One favored readout, especially for PRKs involved in defense, is oxidative burst. Receptors can be expressed in tobacco leaves, and upon addition of ligand, reactive oxygen species (ROS) are produced and can be measured in a luminometer (Chapter 7). This method is particularly useful for the determination of the length and the possible posttranslational modification of peptide ligands.

With new, affordable sequencing technologies, there are vast opportunities to identify and study gene function, including genes encoding ligands and PRKs, in a variety of plant species, to understand plant evolution and diversification, and to transfer knowledge generated in model species to crops. A relevant endeavor is to use bioinformatics, as exemplified in Part III, to identify orthologues of PRKs previously identified in models. This can be cumbersome due to the large number of quite similar receptors that have evolved both through genome and gene duplications and to erroneous or lacking annotations in newly sequenced species. A step-by-step guidance to bioinformatics procedures for the identification of orthologues and improvement of annotations are presented in Chapter 8. From an evolutionary perspective, one can assume that the amino acids of the ectodomain involved in ligand binding suffer from constraints on changes and therefore are highly conserved. This reasoning is the foundation for "conservation mapping" of ectodomains consisting of leucine-rich repeats (LRRs). The LRR receptor-like kinases constitute the largest subgroup of the PRKs, with more than 200 members in *Arabidopsis*. LRR PRKs with known ligands have all more than 20 of these about 24 amino acid-long repeat modules with evenly spaced leucines. The trick with the conservation mapping is to disregard the leucine residues and then to identify the most conserved amino acids. A prerequisite for a meaningful identification of these residues is the identification of the true orthologues of a given receptor in other species. Thus, Chapter 9 exemplifies both the necessary phylogenetic analyses and the use of the mapping program. As a proof of concept, the protocol has been tested successfully on two receptors with solved structure and known ligand and co-receptor binding sites.

Co-receptors may facilitate ligand binding and relay signal transduction, and receptors may also form heterodimers with other receptors. Co-immunoprecipitation is a suitable method for testing interaction between candidate partners identified by genetic or bioinformatics approaches (Chapter 10). A prerequisite is isolation of microsomal membrane fractions from suitable cells or organs (e.g., yeast, tobacco leaves) co-expressing the proteins, often with a tag, and suitable antibodies either directed against the proteins in question or recognizing tags.

Protein-protein interactions may involve the ectodomain and/or the cytoplasmic highly conserved kinase domain. A crucial point is whether a receptor in question is kinase active, i.e., actually capable of phosphorylating itself or other proteins. Testing of autophosphorylation normally requires expression and purification of the full-length receptor or just the kinase domain. Chapter 11 describes wheat germ cell-free production of receptor without extensive purification, in combination with luminescence-based high-throughput assay that can be used for autophosphorylation analysis. Alternatively, in vitro kinase assay can be performed using biotin-streptavidin affinity purification technology.

Having established that a receptor is kinase active, one would like to determine the phosphorylation site and whether the PRK can phosphorylate interacting partners. Approaches to this are found in Chapter 12 and include a protocol for the identification of phosphorylated residues by mass spectrometry as well as a protocol to determine if an interacting partner can be phosphorylated in vitro. The kinase domain of PRKs may also have guanylyl cyclase activity and generate cyclic guanosine monophosphate (cGMP), which lately has been implicated in PRK signaling. The formation of cGMP is detected by mass spectrometry (Chapter 13).

The activity of the kinase domain can be regulated at various levels including posttranslational modification and binding of regulatory proteins, like the calcium-sensing calmodulin (CaM). CaM binding sites are conserved and can be predicted with bioinformatics tools. Chapter 14 provides a detailed protocol for bimolecular fluorescence complementation (BiFC), also called the split-YFP, a method suitable for the investigation of interaction of membrane proteins. The method is based on the assembly of two nonfluorescent halves of a fluorescent protein to a functional fluorophore upon protein interaction. An alternative method is in vitro pull-down assay using heterologously expressed tagged CaM and tagged kinase domain (Chapter 15). The in vitro system allows for controlled reaction conditions such as the concentration of divalent cations. It is further possible to posttranscriptionally modify the receptor kinase prior to the interaction assay.

Although in vitro methods can generate important information, the investigation of the dynamics of PRK complex formation in living plant tissue has become of increasing importance. In vivo fluorescence or Förster resonance energy transfer (FRET) measurements allow a high spatial and temporal resolution for the observation of in vivo changes of plant receptor kinase interactions and complex formation in their natural environment at a subcellular level. Chapter 16 describes in detail how to use FRET techniques for reliable quantitative measurements of in vivo protein interaction. The dynamics of PRKs includes ligand-dependent endocytosis which is conserved across pattern recognition receptors. Confocal microscopy on fluorescently tagged receptors in combination with florescent markers for different subcellular compartments, detailed in Chapter 17, allows the identification of receptor localization along subcellular trafficking routes.

Oslo, Norway *Reidunn Birgitta Aalen*

Contents

Contributors

Reidunn Birgitta Aalen • *Section for Genetics and Evolutionary Biology, Department of Biosciences, University of Oslo, Oslo, Norway*

Ahmed Jawaad Afzal • *Department of Biology, SBA School of Science and Engineering, Lahore University of Management Sciences, Lahore, Pakistan; Department of Plant, Soil and Agricultural Systems, Southern Illinois University, Carbondale, IL, USA*

Moneeza Akbar Agha • *Department of Biology, SBA School of Science and Engineering, Lahore University of Management Sciences, Lahore, Pakistan*

Markus Albert • *Center for Plant Molecular Biology, University of Tübingen, Tübingen, Germany*

María Laura Barberini • *Instituto de Investigaciones en Ingeniería Genética y Biología Molecular, Dr. Hector Torres (INGEBI-CONICET), Buenos Aires, Argentina*

Petra Dietrich • *Molecular Plant Physiology, Department of Biology, University of Erlangen-Nuremberg, Erlangen, Germany*

Cornelia Fischer • *Molecular Plant Physiology, Department of Biology, University of Erlangen-Nuremberg, Erlangen, Germany*

Ursula Fürst • *Center for Plant Molecular Biology, University of Tübingen, Tübingen, Germany*

Chris Gehring • *Division of Biological and Environmental Science and Engineering, 4700 King Abdullah University of Science and Technology, Thuwal, Kingdom of Saudi Arabia*

Christine Kaufmann • *Plant Developmental Biology and Plant Physiology, University of Kiel, Kiel, Germany*

Kohki Kido • *Proteo-Science Center, Ehime University, Matsuyama, Ehime, Japan*

Jin Suk Lee • *Department of Biology, Concordia University, Montreal, QC, Canada*

David Lightfoot • *Department of Plant, Soil and Agricultural Systems, Southern Illinois University, Carbondale, IL, USA*

Jenna Loiseau • *The Sainsbury Laboratory, Norwich Research Park, Norwich, UK*

Yoshikatsu Matsubayashi • *Division of Biological Science, Graduate School of Science, Nagoya University, Nagoya, Japan*

Matthew R. Meyer • *C2N Diagnostics, St. Louis, MO, USA*

Jorge P. Muschietti • *Instituto de Investigaciones en Ingeniería Genética y Biología Molecular, Dr. Hector Torres (INGEBI-CONICET), Buenos Aires, Argentina*

Keiichirou Nemoto • *Proteo-Science Center, Ehime University, Matsuyama, Ehime, Japan*

Shunsuke Nomura • *Proteo-Science Center, Ehime University, Matsuyama, Ehime, Japan*

Akira Nozawa • *Proteo-Science Center, Ehime University, Matsuyama, Ehime, Japan*

Russell J.S. Orr • *Section for Genetics and Evolutionary Biology, Department of Biosciences, University of Oslo, Oslo, Norway*

Misjudeen Raji • *Analytical Core Laboratory, 4700 King Abdullah University of Science and Technology, Thuwal, Kingdom of Saudi Arabia*

A. Gururaj Rao • *Roy J. Carver Department of Biochemistry, Biophysics and Molecular Biology, Iowa State University, Ames, IA, USA*

SILKE ROBATZEK • *The Sainsbury Laboratory, Norwich Research Park, Norwich, UK*
JARKKO SALOJÄRVI • *Division of Plant Biology, Department of Biosciences, Viikki Plant Science Centre, University of Helsinki, Helsinki, Finland*
MARGRET SAUTER • *Plant Developmental Biology and Plant Physiology, University of Kiel, Kiel, Germany*
TATSUYA SAWASAKI • *Proteo-Science Center, Ehime University, Matsuyama, Ehime, Japan*
SHWETA SHAH • *Roy J. Carver Department of Biochemistry, Biophysics and Molecular Biology, Iowa State University, Ames, IA, USA*
HIDEFUMI SHINOHARA • *Division of Biological Science, Graduate School of Science, Nagoya University, Nagoya, Japan*
YVONNE STAHL • *Institute for Developmental Genetics, Heinrich-Heine University, Düsseldorf, Germany*
ALEKSIA VAATTOVAARA • *Division of Plant Biology, Department of Biosciences, Viikki Plant Science Centre, University of Helsinki, Helsinki, Finland*
STEFANIE WEIDTKAMP-PETERS • *Center for Advanced Imaging (CAi), Heinrich-Heine University, Düsseldorf, Germany*
MICHAEL WRZACZEK • *Division of Plant Biology, Department of Biosciences, Viikki Plant Science Centre, University of Helsinki, Helsinki, Finland*
SATOSHI YAMANAKA • *Proteo-Science Center, Ehime University, Matsuyama, Ehime, Japan*

Part I

Expression and Purification of PRKs

Chapter 1

Expression of Plant Receptor Kinases in *E. coli*

Moneeza Akbar Agha, David Lightfoot, and Ahmed Jawaad Afzal

Abstract

Plant receptor kinases play diverse signaling roles in disease resistance and plant development. They represent a large plant gene family with over 600 members in *Arabidopsis thaliana*. While the functions of several members of the receptor kinase family have now been elucidated, a great proportion still remains uncharacterized. The structural and functional characterization of such plant receptor kinases may entail biochemical approaches that require access to purified protein, which can be made possible through heterologous protein expression. This chapter describes a strategy for expressing plant receptor kinases in *E. coli*, a bacterial host that has successfully been used to express and purify certain plant receptor kinase domains, some of which were subsequently used for biochemical assays. As full-length receptor-like kinases may be difficult to express, it is suggested to clone and express domains separately, after having identified domain borders using bioinformatics tools. A detailed cloning protocol is provided, as well as advice for testing expression efficiency and handling of expressed protein ending up in inclusion bodies.

Key words Plant receptor kinases (PRKs), Receptor-like kinases (RLKs), Heterologous protein expression, Leucine-rich repeats (LRRs), Kinase domain

1 Introduction

The ability to detect and respond to changes in the external environment is crucial to the survival of all living organisms. Sessility and the need for photoautotrophic growth place an even greater demand on plants as compared to other living organisms for perceiving and responding to environmental cues [1] as well as external threats. In plants, a diverse group of receptor kinases perform this function by acting as receptors for extracellular signals. They are involved in various biological processes such as developmental regulation [2], defense responses [3], and stress perception [4]. The prototypical plant receptor kinase (*see* Fig. 1) consists of an N-terminal ligand binding extracellular domain, a single membrane spanning region, and an intracellular C-terminal kinase domain involved in downstream signal transduction [5, 6]. Since the cognate ligands for the majority of putative plant receptor

Reidunn Birgitta Aalen (ed.), *Plant Receptor Kinases: Methods and Protocols*, Methods in Molecular Biology, vol. 1621,
DOI 10.1007/978-1-4939-7063-6_1,

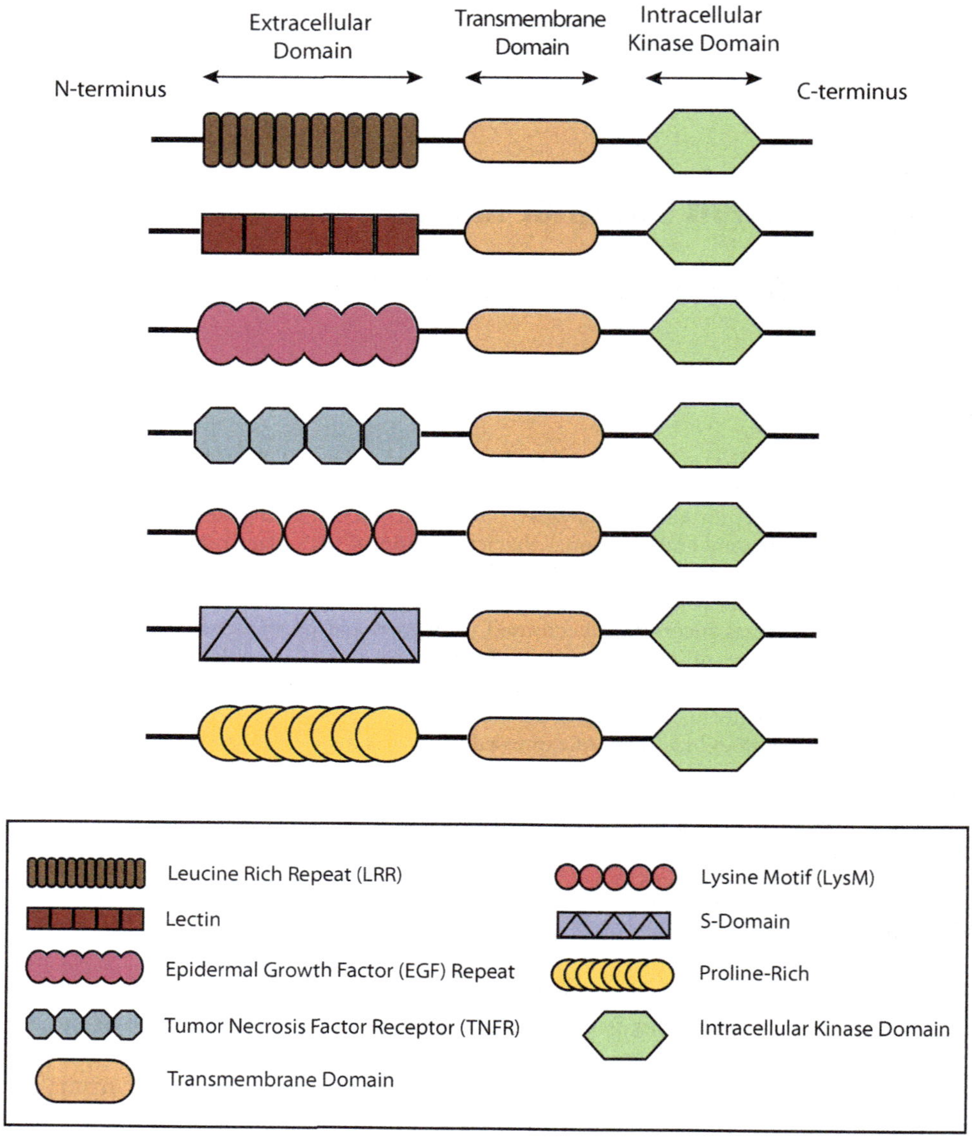

Fig. 1 The domain organization of a typical plant receptor kinase (PRK): A canonical PRK includes a variable N-terminal extracellular domain, a transmembrane domain, and a C-terminal intracellular serine/threonine kinase domain. PRKs have been grouped into 17 subtypes based on the sequence and structural variability of the extracellular domain [3, 6, 9], some of which are depicted in the figure

kinases remain to be identified, they are collectively known as receptor-like kinases (RLKs) [6].

The first RLK was identified in maize [7], and many more have been identified in various plant species since then [8]. The RLK gene family is one of the largest gene families in plants, as opposed to the RLK gene family of animals, which contains far fewer proteins [9]. The *Arabidopsis thaliana* genome is predicted to encode 610 putative RLKs [9], which represents 2.5% of the annotated protein coding region [9] of the genome, while the rice genome contains 1132 RLKs [8]. Additionally, members of the RLK gene family exhibit diverse domain organization and a huge variation in their extracellular domain sequence [3], which confers on them the capacity to perceive an even greater variety of external signals [10]. The large size and the diversity of the plant RLK gene family emphasize its central role in sensing, and transducing, environmental stimuli.

Plant RLKs are classified into 45 different subfamilies based on the characteristics of their extracellular and intracellular kinase domains [10]. LRR-RLKs that contain leucine-rich repeats, which are up to 30 residue long tandem repeats with conserved leucines, constitute the most abundantly represented and well-studied RLK subfamily. However, the functions of many RLKs remain unknown [11], and the nature of their ligands is also not well understood [6]. Among LRR-RLKs, functions for only 30 have so far been ascribed [12]. Moreover, the structural basis of the known ligand–receptor interactions remains to be fully characterized [6]. Structural studies of the intracellular kinase domain can also help obtain a clearer picture of the mechanism by which RLKs activate downstream targets [13]. These questions can be addressed using a combination of genetic, biochemical, structural, and bioinformatics approaches. The application of some of these approaches relies upon the availability of the purified protein kinase. The low expression of some protein kinases [14] makes their isolation from plant tissue an inefficient way of acquiring a high enough yield required for most biochemical assays. Expression and subsequent purification of the receptor kinase from a heterologous expression system present a practical solution to this problem.

E. coli is the most commonly used prokaryotic heterologous expression system owing to the associated ease of genetic manipulation, its short generation time, the relatively low cost of culturing, and a high yield of the recombinant protein product. However, the use of *E. coli* for the expression of eukaryotic proteins has certain limitations, some of which stem from its inability to add eukaryote-specific posttranslational modifications or allow for efficient formation of disulfide bridges. Furthermore, large proteins, particularly membrane proteins, sometimes fail to express at all [15], express to low levels, or form insoluble aggregates upon expression. Additionally, some transmembrane proteins are, on occasion, toxic to the growth of *E. coli* [14].

As relatively large, multi-domain, transmembrane proteins, RLKs might not express at all or not be expressed in a soluble form in *E. coli*, if cloned as full-length proteins. Individually cloning the protein domains contained within the full-length protein increases the likelihood of successful expression and solubility. The intracellular and extracellular domains of the receptor kinase RLK5 [14,

16], the kinase domain of BRI1 [17] and BAK1 [18], the LRR domains of Mi-1 [19], RPM1 [20], and the RLK at RHG1 [21] are among the examples of plant RLK domains successfully expressed in *E. coli*. However, LRR domains tend to form insoluble inclusion bodies upon expression in *E. coli* [21]. While expressing the recombinant protein in a soluble form is preferred, inclusion body formation can sometimes be advantageous, since 80–95% of the inclusion body material is constituted by the target protein [22], which makes them a source of a highly pure protein. The functional target protein can be obtained from inclusion bodies after solubilization and denaturation with a chaotropic agent, such as urea or guanidine hydrochloride, in the presence of a reducing agent, such as β-mercaptoethanol or dithiothreitol (DTT) for cysteine-containing proteins, followed by a refolding procedure [23, 24].

This chapter describes a detailed protocol for cloning and expressing either the full length or a portion of a plant receptor kinase in *E. coli*.

2 Materials

2.1 Bioinformatics Tools

1. InterPro: https://www.ebi.ac.uk/interpro/.
2. SMART: http://smart.embl-heidelberg.de/help/smart_about.shtml.
3. PROSITE: http://prosite.expasy.org/.
4. NCBI Conserved Domain DB: http://www.ncbi.nlm.nih.gov/Structure/cdd/wrpsb.cgi.
5. mGenThreader: http://bioinf.cs.ucl.ac.uk/psipred/.
6. PSI-PRED: http://bioinf.cs.ucl.ac.uk/psipred/.

2.2 PCR Amplification of the Coding Sequence

1. Template DNA.
2. *Pfu/Taq* DNA polymerase and designated buffer.
3. 10 μM forward primer.
4. 10 μM reverse primer.
5. 10 mM dNTPs.
6. Nuclease-free Milli-Q H_2O.

2.3 Gel Elution of the PCR Product

1. Agarose.
2. 0.5 M EDTA pH 8.0: Dissolve 186.1 g of EDTA in 800 mL of distilled H_2O. Adjust the pH to 8.0 with NaOH, and make up the final volume of the solution to 1 L with distilled H_2O.
3. 50× TAE stock solution: Dissolve 242 g of Tris base, 57.1 mL of glacial acetic acid, and 100 mL of 0.5 M EDTA (pH 8.0) in 1 L of distilled H_2O. Store at room temperature.

4. 1× TAE buffer: Add 20 mL of 50× TAE to 800 mL of distilled H_2O.
5. 6× DNA loading buffer: 30% (v/v) glycerol, 0.25% (w/v) bromophenol blue, 0.25% (w/v) xylene cyanol FF.
6. 100 bp or 1 kbp DNA ladder.
7. Ethidium bromide or SYBR Safe.
8. Gel elution kit (e.g., Qiagen's QIAquick Gel Extraction Kit).
9. DNA gel electrophoresis system.

2.4 PCR Product Ligation into pGEMT

1. pGEMT vector with 2× ligation buffer (Promega, Madison, WI, USA).
2. Nuclease-free Milli-Q H_2O.
3. LB media: Dissolve 5 g of tryptone, 2.5 g of yeast extract, and 2.5 g of NaCl in 500 mL of distilled water. Sterilize by autoclaving.
4. LB/ampicillin/IPTG/X-gal plates: Add 5 g of tryptone, 2.5 g of yeast extract, 2.5 g of NaCl, 7.5 g of agar to 500 mL distilled water. Sterilize by autoclaving. Allow the LB agar to cool to 55 °C before adding 500 μL each of filter-sterilized 100 mg/mL ampicillin, 100 mM isopropyl-β-d-galactopyranoside (IPTG) (0.12 g in 5 mL sterile distilled water), and 20 mg/mL X-gal (0.1 g in 1 mL of dimethylformamide) stocks. Pour the LB agar into sterile plastic petri plates.
5. DH5α competent cells.
6. SOC medium: Add 10 g of tryptone, 2.5 g of yeast extract, 1 mL of 5 M NaCl, 1.25 mL of 1 M KCl, 5 mL of 1 M $MgCl_2$, 5 mL of 1 M $MgSO_4$ to 375 mL of distilled water. Adjust the pH of the solution to 7 using 5 M NaOH. Bring the volume to 490 mL with distilled H_2O. Sterilize by autoclaving. Add 10 mL of filter-sterilized 1 M glucose to the media.
7. QIAprep Spin Miniprep Kit (Qiagen, Hilden, Germany).

2.5 Colony PCR

1. *Taq* DNA polymerase and designated buffer.
2. 10 μM each of relevant forward and reverse primers.
3. 10 mM dNTPs.
4. Nuclease-free Milli-Q H_2O.
5. Template: Pick part of each colony to be screened and dissolve it in 50 μL distilled water, heat the solution at 95 °C, and centrifuge at 12000 × *g* for 5 min. Use 5 μL of the supernatant as the template in a 50 μL reaction.
6. Sterile toothpicks.
7. Centrifuge with speeds ≥12000 × *g*.

2.6 Restriction Digestion of Insert and Recipient Expression Vector

1. Appropriate restriction enzymes with required buffers.
2. Nuclease-free Milli-Q H_2O.
3. Water bath set at the required temperature.
4. Materials required for gel elution.

2.7 Ligation of the Insert with the Vector

1. Expression vector.
2. T4 DNA ligase (NEB, Ipswich, MA, USA).
3. 10× T4 DNA ligase reaction buffer (NEB, Ipswich, MA, USA).
4. Nuclease-free Milli-Q H_2O.
5. Water bath set at 65 °C.

2.8 Transformation of the Ligation Reaction

1. LB media.
2. Specific antibiotic.
3. LB agar containing the required selection antibiotic.
4. DH5α competent cells.
5. SOC media.
6. QIAprep Spin Miniprep Kit (Qiagen, Hilden, Germany).
7. Temperature-controlled shaking incubator.

2.9 Induction of Recombinant Protein Expression

1. LB media and agar supplemented with the required antibiotic.
2. Specific antibiotic (kanamycin or ampicillin for the pET vector system).
3. BL21 (DE3) or BL21 CodonPlus™ (DE3)-RIL or BL21 (DE3) CodonPlus-RP.
4. 1 M isopropyl-β-d-galactopyranoside (IPTG).
5. Spectrophotometer.

2.10 Protein Extraction

1. BugBuster® 10× Protein Extraction Reagent or BugBuster® Protein Extraction Reagent.
2. Lysonase™ Bioprocessing Reagent.
3. Protease inhibitor cocktail: 100× (EDTA-free).
4. Tris–HCl: 50 mM pH 7.5.
5. Sodium chloride (NaCl): 150 mM.
6. β-mercaptoethanol: 20 mM.
7. Phosphate buffer for inclusion bodies: 20 mM sodium phosphate pH 7.8, 500 mM NaCl, and 20 mM β-mercaptoethanol.
8. Tris-based buffer for inclusion bodies: 50 mM Tris–HCl pH 7.5, 150 mM NaCl, and 20 mM β-mercaptoethanol.
9. 6.0–8.0 M urea.
10. 4.0–6.0 M guanidine hydrochloride.

2.11 SDS-PAGE

1. 40% (w/v) acrylamide/bisacrylamide (cross-linker ratio 37:5:1).
2. 10% (w/v) ammonium persulfate (APS).
3. Tetramethylethylenediamine (TEMED).
4. 1.5 M Tris (pH 8.8): Dissolve 181.7 g of Tris base in 800 mL of distilled H_2O. Adjust the pH to 8.8 with concentrated HCl. Adjust the volume to 1 L. Sterilize by autoclaving.
5. 1 M Tris (pH 6.8): Dissolve 12.1 g of Tris base in 80 mL of distilled H_2O. Adjust the pH to 6.8 with concentrated HCl. Adjust the volume to 100 mL. Sterilize by autoclaving. Store at room temperature.
6. 5× SDS loading buffer: 50% glycerol, 0.25 M Tris–HCl pH 6.8, 10% SDS, 0.5% bromophenol blue, and 100 mM DTT. DTT must be added to the buffer just prior to use.
7. Protein ladder/marker.
8. 10× running buffer stock: Dissolve 30.3 g of Tris base, 144 g of glycine, and 10 g of SDS in 1 L of distilled H_2O.
9. 1× running buffer: Add 100 mL of 10× running buffer to 900 mL of distilled H_2O.
10. Coomassie Blue R-250 staining solution: 0.1% (w/v) Coomassie Brilliant Blue R-250, 50% (v/v) methanol, 10% (v/v) glacial acetic acid, and 40% (v/v) distilled H_2O.
11. Coomassie destaining solution: 40% (v/v) methanol, 10% (v/v) acetic acid.
12. Protein gel electrophoresis system (BioRad Mini-Protean vertical electrophoresis cell or equivalent).

2.12 Western Blotting

1. 10× transfer buffer stock: Dissolve 30.3 g of Tris base and 144 g of glycine in 1 L of distilled H_2O.
2. 1× transfer buffer: Add 100 mL of 10× transfer buffer and 200 mL of methanol to 700 mL of distilled H_2O.
3. 10× TBS (Tris-buffered saline): Dissolve 24.2 g of Tris base and 87.7 g of NaCl in 900 mL of distilled H_2O. Adjust the pH to 7.6 with HCl, and bring the final volume to 1 L.
4. 1× TBST (Tris-buffered saline, 0.1% Tween 20): Add 1 mL of Tween 20 to 100 mL of 10× TBS to 899 mL of distilled H_2O.
5. Ponceau S stain: 0.1 g of Ponceau S stain, 5 mL of acetic acid, and 95 mL of distilled H_2O.
6. Blocking solution: 3% (w/v) Bovine Serum Albumin (BSA) or 5% (w/v) Skimmed milk in TBST.
7. Primary antibody.
8. HRP-conjugated secondary antibody.
9. ECL Western blotting detection reagents.
10. The ChemiDoc XRS+ system or an equivalent imaging system.

3 Methods

3.1 Bioinformatics Analysis

1. Cloning portions of the RLK as separate fragments instead of the full-length protein can provide the opportunity of excluding potentially hydrophobic, destabilizing and/or disordered regions, thereby increasing the likelihood of generating a soluble, functional, and stable recombinant protein. The full-length protein can be split into fragments based on predicted domain boundaries (*see* Fig. 1 and **Note 1**).
2. For conserved domains, the core domain region can be identified by employing the InterProScan tool [25], which combines information from 11 different protein databases that classify domains based on sequence homology. A structure-based sequence alignment of the gene of interest with proteins of known structure can be obtained using the fold recognition tool GenThreader [26] for further refinement of domain boundaries.
3. Domain boundaries should be demarcated such that they do not interrupt any predicted secondary structural elements. Additionally, domains should not have hydrophobic residues and low complexity regions at their termini [27]. Secondary structure predictions can be obtained using PSIPRED [28].

3.2 Cloning the Expression Construct

1. Once the region of protein to be expressed has been selected, the next step is to pick an appropriate expression vector. While choosing an expression vector, the plasmid characteristics to be considered include the type of replicon which dictates the plasmid copy number per cell, the kind of the promoter that will control the expression of the gene of interest, the antibiotic selection marker present, the restriction sites contained within the multiple cloning site (MCS) of the vector, and the kind of fusion tag (*see* **Note 2**) to be added to the protein and whether it is to be attached at the N- or the C-terminus. The availability of a proteolytic cleavage site within the vector (*see* **Note 3**) for the subsequent removal of the fusion tag is another factor that must be taken into account. This protocol describes the use of frequently employed vectors from the pET expression system that allow IPTG-inducible expression of the recombinant protein of interest by bacteriophage T7 RNA polymerase.
2. Prepare the insert consisting of the selected protein coding sequence that will be ligated into the expression vector between two restriction sites chosen (*see* **Note 4**) from the several contained within the MCS of the vector after excluding the possibility of their occurrence within the coding sequence of your gene (*see* **Note 5**).
   (a) Design primers (*see* **Note 6**) for the PCR amplification of the cDNA corresponding to the chosen protein portion

from an expressed sequence tag clone or total cDNA synthesized using mRNA obtained from plant tissue known to express the RLK of interest. The 5′ end of the forward primer must start with a 3–10 nucleotide (*see* **Note 7**) linker sequence followed by the first restriction site selected from the MCS of the expression vector. Add a 12- to 18-nucleotide sequence stretch corresponding to the 5′ end of the coding sequence to be amplified after the restriction site at the 3′ end of the forward primer. Add the start codon ATG before the gene sequence if the entire coding sequence is not being cloned and no fusion tag is being added to the recombinant protein. The 5′ end of the reverse primer must begin with a 3–10 nucleotide linker followed by the second restriction site picked from the MCS and then include the reverse complement to the terminal 12–18 bp of the selected protein coding sequence. The stop codon TAA must be included after the restriction site in case an N-terminal fusion tag or no fusion tag is being added to the recombinant protein.

(b) Set up a 50 μL PCR reaction with *Taq* DNA polymerase (*see* **Note 8**) following the manufacturer's instructions.

(c) Analyze the PCR product by gel electrophoresis using a 0.8–2% (w/v) TAE agarose gel, depending upon the size of the PCR product. Elute the product from the gel using a gel extraction kit as per the manufacturer's instructions. Alternatively, purify the PCR product using a PCR clean up kit (*see* **Note 9**). Measure the concentration of the eluted or purified PCR product.

(d) The eluted PCR product can be restriction digested and directly cloned into the chosen expression vector. Alternatively, the eluted PCR product can be cloned into a cloning vector to facilitate subsequent subcloning into several expression vectors, which is the approach described here (*see* **Note 10**).

(e) Set up a 10 μL ligation reaction between the eluted PCR product and the pGEMT vector using T4 DNA ligase and the 2× ligation buffer (Promega, Madison, WI, USA) according to the manufacturer's instructions (*see* **Note 11**). Incubate the reaction for 1 h at room temperature. Transform 2 μL of the ligation reaction mixture into 50 μL of chemically competent DH5α cells using the standard heat shock transformation protocol. Add 900 μL of SOC media to the mix and incubate for 1.5 h at 37 °C with shaking at 250 rpm. Plate 100–500 μL of the transformation culture onto LB/ampicillin/IPTG/X-gal plates, and incubate the plates overnight at 37 °C. Select positive (white) colonies and use part of each colony to screen for insert presence by PCR (*see* **Note 12**), and grow the rest overnight in 5 mL ampicillin supplemented LB media at 37 °C. Isolate recombinant plasmid DNA using a plasmid miniprep kit following the manufacturer's protocol. Perform a diagnostic restriction digestion to confirm the size and orientation of the insert. Analyze the recombinant plasmid by sequencing to exclude the possibility of mutations within the insert.

(f) Set up a 50 μL double digestion reaction of 10 μg of the recombinant pGEMT vector with 10–20 units (depending upon the enzyme) each of the two restriction enzymes and a compatible buffer (*see* **Note 13**) at the optimum temperature of the enzymes to release the insert-carrying terminal sticky ends for ligation into the expression vector. Analyze the restriction reaction by agarose gel electrophoresis. Elute the desired fragment using a gel elution kit. Determine the concentration of the eluted DNA.

3. Digest 10 μg of the expression vector with the two appropriate restriction enzymes and a compatible buffer in a 50 μL reaction at the recommended temperature to produce sticky ends compatible to those of the insert. Resolve the restriction reaction on an agarose gel, and elute the fragment corresponding to digested DNA using a gel elution kit.
4. Set up a 20 μL ligation reaction with 1 μL T4 DNA ligase, 2 μL of 10× T4 DNA ligase reaction buffer, 0.020 pmol (50–100 ng) of vector DNA, and 0.060 pmol of insert DNA (*see* **Note 14**). Use nuclease-free Milli-Q H_2O to bring the reaction to volume. Incubate the reaction at 16 °C overnight. Heat inactivate the reaction at 65 °C for 10 min. Transfer the reaction to ice. Transform 1–5 μL (100 ng of DNA) of the ligation reaction into 50 μL chemically competent DH5α cells. Add 250 μL of SOC medium to the transformation reaction, and incubate at 37 °C in a shaking incubator set at 200 rpm for 1 h. Plate 100–500 μL of the transformation culture on LB agar plates supplemented with the required selection antibiotic (kanamycin or ampicillin for a pET vector), and incubate overnight at 37 °C. Select 5–10 colonies, and screen them by PCR to confirm the presence of the gene of interest. Part of the colony may be used for the PCR while the remaining must be grown in 5 mL of antibiotic-containing LB media at 37 °C overnight. Use a plasmid miniprep kit to isolate recombinant plasmid DNA from the culture of the colonies that produced a positive PCR result. Sequence the isolated construct to confirm in frame cloning.

3.3 IPTG Induction of Recombinant Protein Expression

Genes inserted into expression vectors containing IPTG-inducible promoters can be expressed using the following protocol:

1. Transform 50 ng of the recombinant expression plasmid into 50 μL of a compatible chemically competent *E. coli* expression host strain. BL21 (DE3) (*see* **Note 15**) can be used if the expression vector under use belongs to the pET vector system. Plate the transformation culture on an LB antibiotic selection plate, and incubate it overnight at 37 °C.
2. Pick a single colony from the plate (*see* **Note 16**), and use it to inoculate 3 mL of antibiotic supplemented LB media in a 15 mL culture tube. Incubate the LB culture overnight (*see* **Note 17**) at 30 °C in a shaking incubator set at 250 rpm.
3. Inoculate 10 mL of antibiotic-supplemented LB with 100 μL of overnight culture in a 50 mL conical tube. Incubate the culture at 37 °C, and monitor its OD_{600} until it reaches 0.5–0.7 (mid log phase). At this point, transfer 5 mL each of culture into two 50 mL conical tubes. Add IPTG to a final concentration of 1 mM (*see* **Note 18**) to one of the two 5 mL culture tubes. The other culture will serve as the uninduced control. Incubate the cultures at 37 °C (*see* **Note 18**).
4. Collect a sample from the uninduced and induced cultures at 2, 3, 4, and 6 h after IPTG induction. At the time of collection, dilute each culture to an OD_{600} of 0.8 using LB media to enable evaluation of protein expression across the different time points from a constant number of bacterial cells. Transfer 1 mL of the OD_{600} adjusted culture to a 1.5 mL microcentrifuge tube.
5. Split the 1 mL collection into two portions of 800 and 200 μL each.
6. The 200 μL culture will be used to assess total protein expression and the 800 μL culture to evaluate protein solubility.

3.4 Analyzing Total Recombinant Protein Expression

1. Centrifuge each 200 μL culture at maximum speed for 5 min at 4 °C, decant the supernatant. Resuspend the cell pellet in 100 μL of 1× SDS loading buffer.
2. Heat the samples at 95 °C for 10 min. Analyze the protein samples by SDS-PAGE (*see* **Note 19**). Carry out gel electrophoresis at 5–10 V/cm for 10–15 min and then at 10–15 V/cm for 1–2 h until the dye front reaches the bottom of the gel. Stain the gel with Coomassie Brilliant Blue solution. Selective protein enrichment in the induced sample as compared to the uninduced sample at the expected MW of the recombinant protein indicates successful expression.
3. Identify the time point after induction exhibiting the highest protein expression for future large-scale induction experiments.

3.5 Analyzing Recombinant Protein Solubility

1. Centrifuge each 800 μL sample at 6000 × *g* for 20 min at 4 °C, decant the supernatant. Store the cell pellet at −20 °C or proceed to the next step.
2. Resuspend the cell pellet in 160 μL (one fifth of the culture volume) of BugBuster® Protein Extraction Reagent (*see* **Note 20**), and incubate on a rotating mixer at slow speed for 10–20 min at room temperature. Centrifuge at 16,000 × *g* for 20 min at 4 °C. Carefully transfer the supernatant (*see* **Note 21**) into a fresh microcentrifuge tube, and add 40 μL of 5× SDS loading buffer. This is the soluble fraction.
3. Resuspend the cell pellet in 200 μL of 1× SDS loading buffer to obtain the insoluble fraction. Heat both the soluble and insoluble fractions obtained from the induced and uninduced cultures at 95 °C for 5 min.
4. Load 10–15 μL of the protein samples on a 10% SDS-PAGE gel (*see* **Note 19**) to analyze the expression and solubility of the recombinant protein. Preform gel electrophoresis according to the standard protocol. Stain the gel with Coomassie Brilliant Blue solution. Compare the protein levels between the soluble and insoluble fractions of the induced and uninduced cultures at the expected MW of the recombinant protein. For a soluble protein, greater amount of protein should be present in the soluble fraction, while the opposite would be true for an insoluble protein (*see* **Notes 22** and **23**).

3.6 Verification of RLK Expression by Western Blotting

1. The expression of the target recombinant protein can be verified by Western blotting using a target-specific antibody or a fusion tag-specific antibody if a fusion tag was added to the recombinant protein.
2. Load the protein samples on a 10% SDS-PAGE gel along with a protein marker. Perform gel electrophoresis at 5–10 V/cm for 10–15 min and then at 10–15 V/cm for 1–2 h until the dye front reaches the bottom of the gel. Disassemble the setup, remove the gel, rinse it with distilled H_2O, and soak it in 1× transfer buffer for 5 min.
3. Assemble the transfer sandwich with the gel facing the anode and a membrane cut to the dimensions of the gel facing the cathode. Either a nitrocellulose or a methanol activated PVDF membrane can be used. Place the transfer cassette in the tank, and transfer at 100 V for 60–90 min in the cold room. Transfer of proteins to the membrane can be checked using Ponceau S staining.
4. Block the membrane in 10–15 mL of blocking buffer for 1 h at room temperature. This volume is adequate for a 9 × 6 cm^2 membrane. The volume may be increased for larger membranes.

5. Wash the membrane three times for 5–10 min each with 10–15 mL TBST.
6. Incubate the membrane in 10 mL of primary antibody prepared at the dilution recommended by the manufacturer overnight at 4 °C.
7. Rinse the membrane with 10 mL of TBST, and wash it three times for 10 min each with 15 mL of TBST.
8. Incubate the membrane with a species-appropriate HRP-conjugated secondary antibody diluted according to the manufacturer's instructions for 1 h at room temperature.
9. Wash three times for 15 min each with 15 mL of TBST.
10. Probe the blot using an ECL detection reagent.
11. Use the ChemiDoc XRS+ system in chemiluminescence mode to detect the signal from the recombinant protein.

4 Notes

1. An accurate delineation of domain boundaries is crucial to the process, since an underestimation of domain length may result in a buried/charged N- or C-terminus, while an overestimation might introduce structurally disordered regions that can lead to protein aggregation [29].
2. A fusion tag can either consist of an epitope containing only a few residues or a full polypeptide. It is advantageous to include a fusion tag in your expression construct for multiple reasons. The addition of epitope tags enables detection and affinity-based purification, while polypeptide tags offer the added advantage of increasing recombinant protein solubility [30]. Polypeptide tags such as maltose-binding protein (MBP), glutathione S-transferase (GST), thioredoxin (TRX), N-utilization substance protein A (NusA), and small ubiquitin-related modifier (SUMO) are some examples of available solubility enhancer tags [31]. A comparison between MBP, GST, and TRX has identified MBP as the most effective solubilizing agent among the three [32]. The most commonly used small peptide tags include FLAG-, poly-His-, c-myc-, poly-Arg-, S-, and Strep II- [33]. The availability of commercial antibodies against all of these tags facilitates not just the detection of the recombinant protein by Western blotting [30] but also by certain biochemical assays such as co-immunoprecipitation. Poly-His-tagged proteins can be purified by employing immobilized metal ion chromatography (IMAC) using Ni^{2+}-coated NTA resin, while GST-tagged proteins and MBP-tagged proteins can be affinity purified using glutathione and amylose-coated agarose resin, respectively [30]. Other epitope-tagged proteins can be purified with the use of antibody-based chromatography [34].

3. The fusion tag needs to be proteolytically removed for some applications (such as antibody production and structural studies). This can be achieved either by using a vector offering a protease cleavage site between the fusion tag and the target protein or by engineering the protease cleavage site within the forward or the reverse primer (between the restriction site and the gene sequence) under use for PCR amplification of the coding sequence.
4. It is advisable to choose two restriction sites that are at least 12 bp apart within the MCS to allow for efficient restriction. If the distance between the two chosen restriction sites is less than 12 bp, the expression vector may have to be sequentially digested prior to ligation starting with the less efficient restriction enzyme.
5. If none of the restriction sites present within the MCS can be utilized, an isocaudomer with a different recognition site that will produce compatible sticky ends on the insert may be used instead. The restriction site of the isocaudomer would have to be added to the primers meant for PCR amplifying the gene sequence in that case. Alternatively, annealed oligomers (adaptors) may be used to introduce the required restriction sites into the MCS.
6. While designing primers, ensure that the melting temperature (T_m) of the sequence corresponding to the gene in the forward and reverse primer is between 60 and 70 °C. The T_m of the forward and the reverse primers must be within 1–2 °C of each other for simpler optimization of PCR conditions. Ensure that there is little to no complementarity between the forward and the reverse primers to prevent primer dimer formation, which can significantly lower PCR efficiency. Primers may be designed and analyzed using softwares such as the Vector NTI or Serial Cloner.
7. Addition of extra bases prior to the restriction sites allows for efficient restriction of the PCR product if it is to be directly cloned into an expression vector. The number of extra bases that need to be added depends upon the restriction enzyme.
8. A proofreading DNA polymerase may be used to generate the PCR product instead of *Taq* polymerase, especially while amplifying a larger insert, when there is greater risk of introducing mutations. To adapt the purified PCR product for TA cloning, 3′-A tails may be added later in an A-tailing reaction that uses *Taq* polymerase. Alternatively, a mixture of *Pfu* and *Taq* can be used to carry out the PCR reaction.
9. The PCR product must be purified by gel electrophoresis if multiple bands of amplified DNA are observed on the agarose gel indicating nonspecific amplification. A PCR purification kit

may be used when you obtain a single band upon analyzing the PCR product on a TAE agarose gel.

10. A *Taq* polymerase-generated PCR product carries 3′-adenine overhangs that can anneal with the 3′-thymidine overhangs of a linearized T vector such as pGEMT to allow for successful ligation.
11. A 1:1 insert/vector molar ratio is recommended for TA cloning. Different insert/vector molar ratios might have to be tried if the ligation reaction with a 1:1 insert/vector molar ratio is unsuccessful.
12. The insert-specific forward or reverse primer can be used in conjunction with a vector-specific primer during a colony PCR to assess the orientation of the insert in the cloning vector.
13. The recombinant pGEMT vector may be digested sequentially if the two restriction enzymes do not have a compatible buffer, if one of the enzymes is a poor cutter, or if the enzymes have different optimum temperatures.
14. An insert/vector molar ratio of 3:1 or 2:1 is recommended for successful cloning. Use this formula to calculate the amount of insert and vector to be included in the ligation reaction: ((ng of vector × kb size of insert)/kb size of vector) × (insert:vector molar ratio)) = ng of insert.
15. A mismatch between the codon usage of the gene of interest and that of the *E. coli* expression host is detrimental for heterologous protein expression [35]. This problem can be addressed by using a codon bias-adjusted *E. coli* strain such as BL21 CodonPlus (DE3)-RIL or BL21 (DE3) CodonPlus-RP that encodes certain genes for low-abundance *E. coli* tRNAs. C41 (DE3) and C43 (DE3) have been reported to produce higher yields for membrane proteins [36]. These may be used when cloning either the full-length receptor kinase or the membrane-spanning domain.
16. Recombinant protein expression might at times differ from clone to clone. Multiple colonies may be assayed for protein expression, and the one expressing the recombinant protein to high levels can be preserved in the form of a glycerol stock for large-scale purification experiments.
17. Ampicillin is degraded by the β-lactamase that is secreted by the bacterial cells that have incorporated the expression vector carrying the marker for ampicillin resistance [37]. Antibiotic degradation will eliminate the selective pressure, thereby allowing the bacterial cells not containing the expression vector to grow. Growing the culture at higher temperatures will lead to faster growth of resistant bacterial cells and hence faster degradation of ampicillin. For that reason, the overnight culture

must not be set at temperatures above 30 °C while using ampicillin as the selection antibiotic. In order to remove β-lactamase from the overnight culture, it must be centrifuged, pelleted, and resuspended in fresh media before using it for the inoculation of the induction culture.

18. The optimum IPTG concentration, induction temperature, and duration will vary, depending upon the recombinant protein of interest and the expression vector under use. IPTG concentrations between 0.05 to 5 mM and induction temperatures ranging from 15 to 37 °C can be tried to optimize the expression conditions of the recombinant protein.
19. A higher or lower percentage of resolving gel may need to be used, for optimal resolution, depending upon the molecular weight of the recombinant protein.
20. BugBuster® Protein Extraction Reagent consists of a Tris-based buffer system. If the recombinant protein necessitates the use of an alternative buffer system (such as phosphate or PIPPS based), BugBuster® 10× Protein Extraction Reagent may instead be diluted in the chosen buffer. However, BugBuster® 10× Protein Extraction Reagent is compatible only with basic buffer systems and cannot be used with buffers below a pH of 5. Protein yield may be increased by supplementing BugBuster® Protein Extraction Reagent with Lysonase™ Bioprocessing Reagent that consists of a mixture of the endonuclease Benzonase and Lysozyme. The former hydrolyzes nucleic acids to reduce lysate viscosity, while the latter degrades the bacterial cell wall, there by promoting protein extraction. Additionally, 1 μL of 100× protease inhibitor cocktail per 100 μL of the protein extraction reagent can be added to prevent the proteolytic degradation of proteins. An EDTA-free cocktail must be used for proteins that are to be purified using IMAC.
21. Prevent the contamination of the supernatant with the pellet to avoid overestimating the amount of protein present in the soluble fraction.
22. Lowering the induction temperature (an overnight induction at 25 °C might be a good starting point), the IPTG concentration, and the duration of induction and changing the media composition are all among the approaches that may help to enhance the solubility of a recombinant protein, which at the first induction attempt seems to be accumulating in inclusion bodies. If these strategies fail to produce a soluble recombinant protein, then it can be purified from inclusion bodies under denaturing conditions and subsequently renatured.
23. Proteins accumulating in the insoluble fraction as inclusion bodies can be solubilized either in a phosphate or Tris-based

buffer containing 6.0–8.0 M urea or 4.0–6.0 M guanidine hydrochloride at 25 °C for 60 min. The optimal solubilization conditions (e.g., the compatibility of the buffer system as well as the concentration of salt, the reducing agent, and the chaotrope needed) would vary depending on the recombinant protein and must first be identified by testing small-scale cultures (1–3 mL) before large-scale experiments can be set up. After solubilization, centrifuge the solution at 10,000 × *g* for 30 min at 4 °C to get rid of any residual insoluble aggregates. The extracted protein may then either be purified under denaturing conditions or refolded prior to purification.

References

1. McCarty DR, Chory J (2000) Conservation and innovation in plant signaling pathways. Cell 103(2):201–209. doi:10.1016/S0092-8674(00)00113-6
2. Becraft PW (2002) Receptor kinase signaling in plant development. Annu Rev Cell Dev Biol 18:163–192. doi:10.1146/annurev.cellbio.18.012502.083431
3. Shiu SH, Bleecker AB (2001) Plant receptor-like kinase gene family: diversity, function, and signaling. Sci STKE 2001(113):re22. doi:10.1126/stke.2001.113.re22
4. Osakabe Y, Yamaguchi-Shinozaki K, Shinozaki K, Tran L-SP (2013) Sensing the environment: key roles of membrane-localized kinases in plant perception and response to abiotic stress. J Exp Bot 64(2):445–458. doi:10.1093/jxb/ers354
5. Tichtinsky G, Vanoosthuyse V, Cock JM, Gaude T (2003) Making inroads into plant receptor kinase signalling pathways. Trends Plant Sci 8(5):231–237. doi:10.1016/S1360-1385(03)00062-1
6. Torii KU (2009) Transmembrane receptors in plants: receptor kinases and their ligands. In: Annual plant reviews volume 33: intracellular signaling in plants. Wiley-Blackwell, pp 1–29. doi:10.1002/9781444302387.ch1
7. Walker JC, Zhang R (1990) Relationship of a putative receptor protein kinase from maize to the S-locus glycoproteins of Brassica. Nature 345(6277):743–746. doi:10.1038/345743a0
8. Shiu S-H, Karlowski WM, Pan R, Tzeng Y-H, Mayer KFX, Li W-H (2004) Comparative analysis of the receptor-like kinase family in Arabidopsis and rice. Plant Cell 16(5):1220–1234. doi:10.1105/tpc.020834
9. Shiu S-H, Bleecker AB (2001) Receptor-like kinases from Arabidopsis form a monophyletic gene family related to animal receptor kinases. Proc Natl Acad Sci USA 98(19):10763–10768. doi:10.1073/pnas.181141598
10. Shiu SH, Bleecker AB (2003) Expansion of the receptor-like kinase/Pelle gene family and receptor-like proteins in Arabidopsis. Plant Physiol 132(2):530–543. doi:10.1104/pp.103.021964
11. Fischer I, Dievart A, Droc G, Dufayard JF, Chantret N (2016) Evolutionary dynamics of the leucine-rich repeat receptor-like kinase (LRR-RLK) subfamily in angiosperms. Plant Physiol 170(3):1595–1610. doi:10.1104/pp.15.01470
12. Gou X, He K, Yang H, Yuan T, Lin H, Clouse SD, Li J (2010) Genome-wide cloning and sequence analysis of leucine-rich repeat receptor-like protein kinase genes in Arabidopsis thaliana. BMC Genomics 11(1):1–15. doi:10.1186/1471-2164-11-19
13. Han Z, Sun Y, Chai J (2014) Structural insight into the activation of plant receptor kinases. Curr Opin Plant Biol 20:55–63. doi:10.1016/j.pbi.2014.04.008
14. Horn MA, Walker JC (1995) Chapter 37 Expression and assay of autophosphorylation of recombinant protein kinases. In: Galbraith DW, Bohnert HJ, Bourque PB (eds) Methods in cell biology, vol 49. Academic Press, pp 531–541. doi:10.1016/S0091-679X(08)61478-8
15. Peti W, Page R (2007) Strategies to maximize heterologous protein expression in *Escherichia coli* with minimal cost. Protein Expr Purif 51(1):1–10. doi:10.1016/j.pep.2006.06.024
16. Horn MA, Walker JC (1994) Biochemical properties of the autophosphorylation of RLK5, a receptor-like protein kinase from Arabidopsis thaliana. Biochim Biophys Acta 1208(1):65–74
17. Oh M-H, Ray WK, Huber SC, Asara JM, Gage DA, Clouse SD (2000) Recombinant brassinosteroid insensitive 1 receptor-like kinase autophosphorylates on serine and threonine residues and phosphorylates a conserved peptide motif in vitro. Plant Physiol 124(2):751–766
18. Li J, Wen J, Lease KA, Doke JT, Tax FE, Walker JC (2002) BAK1, an Arabidopsis LRR

receptor-like protein kinase, interacts with BRI1 and modulates brassinosteroid signaling. Cell 110(2):213–222. doi:10.1016/S0092-8674(02)00812-7

19. Tameling WI, Elzinga SD, Darmin PS, Vossen JH, Takken FL, Haring MA, Cornelissen BJ (2002) The tomato R gene products I-2 and MI-1 are functional ATP binding proteins with ATPase activity. Plant Cell 14(11):2929–2939
20. Boyes DC, Nam J, Dangl JL (1998) The Arabidopsis thaliana RPM1 disease resistance gene product is a peripheral plasma membrane protein that is degraded coincident with the hypersensitive response. Proc Natl Acad Sci USA 95(26):15849–15854
21. Afzal AJ, Lightfoot DA (2007) Soybean disease resistance protein RHG1-LRR domain expressed, purified and refolded from *Escherichia coli* inclusion bodies: preparation for a functional analysis. Protein Expr Purif 53(2):346–355. doi:10.1016/j.pep.2006.12.017
22. Baneyx F, Mujacic M (2004) Recombinant protein folding and misfolding in *Escherichia coli*. Nat Biotechnol 22(11):1399–1408. doi:10.1038/nbt1029
23. Fischer B, Perry B, Sumner I, Goodenough P (1992) A novel sequential procedure to enhance the renaturation of recombinant protein from *Escherichia coli* inclusion bodies. Protein Eng 5(6):593–596
24. Misawa S, Kumagai I (1999) Refolding of therapeutic proteins produced in *Escherichia coli* as inclusion bodies. Biopolymers 51(4):297–307. doi:10.1002/(sici)1097-0282(1999)51:4<297::aid-bip5>3.0.co;2-i
25. Mitchell A, Chang H-Y, Daugherty L, Fraser M, Hunter S, Lopez R, McAnulla C, McMenamin C, Nuka G, Pesseat S, Sangrador-Vegas A, Scheremetjew M, Rato C, Yong S-Y, Bateman A, Punta M, Attwood TK, Sigrist CJA, Redaschi N, Rivoire C, Xenarios I, Kahn D, Guyot D, Bork P, Letunic I, Gough J, Oates M, Haft D, Huang H, Natale DA, Wu CH, Orengo C, Sillitoe I, Mi H, Thomas PD, Finn RD (2015) The InterPro protein families database: the classification resource after 15 years. Nucleic Acids Res 43(Database issue):D213–D221. doi:10.1093/nar/gku1243
26. Lobley A, Sadowski MI, Jones DT (2009) pGenTHREADER and pDomTHREADER: new methods for improved protein fold recognition and superfamily discrimination. Bioinformatics 25(14):1761–1767. doi:10.1093/bioinformatics/btp302
27. Structural Genomics C, Architecture et Fonction des Macromolécules B, Berkeley Structural Genomics C, China Structural Genomics C, Integrated Center for S, Function I, Israel Structural Proteomics C, Joint Center for Structural G, Midwest Center for Structural G, New York Structural Genomi XRCfSG, Northeast Structural Genomics C, Oxford Protein Production F, Protein Sample Production Facility MDCfMM, Initiative RSGP, Complexes S (2008) Protein production and purification. Nat Methods 5(2):135–146. doi:10.1038/nmeth.f.202
28. McGuffin LJ, Bryson K, Jones DT (2000) The PSIPRED protein structure prediction server. Bioinformatics 16(4):404–405. doi:10.1093/bioinformatics/16.4.404
29. Reich S, Puckey LH, Cheetham CL, Harris R, Ali AA, Bhattacharyya U, Maclagan K, Powell KA, Prodromou C, Pearl LH, Driscoll PC, Savva R (2006) Combinatorial domain hunting: an effective approach for the identification of soluble protein domains adaptable to high-throughput applications. Protein Sci 15(10):2356–2365. doi:10.1110/ps.062082606
30. Rosano GL, Ceccarelli EA (2014) Recombinant protein expression in *Escherichia coli*: advances and challenges. Front Microbiol 5:172. doi:10.3389/fmicb.2014.00172
31. Costa S, Almeida A, Castro A, Domingues L (2014) Fusion tags for protein solubility, purification and immunogenicity in *Escherichia coli*: the novel Fh8 system. Front Microbiol 5:63. doi:10.3389/fmicb.2014.00063
32. Kapust RB, Waugh DS (1999) *Escherichia coli* maltose-binding protein is uncommonly effective at promoting the solubility of polypeptides to which it is fused. Protein Sci 8(8):1668–1674. doi:10.1110/ps.8.8.1668
33. Terpe K (2003) Overview of tag protein fusions: from molecular and biochemical fundamentals to commercial systems. Appl Microbiol Biotechnol 60(5):523–533. doi:10.1007/s00253-002-1158-6
34. Kimple ME, Brill AL, Pasker RL (2013) Overview of affinity tags for protein purification. Curr Protoc Protein Sci. doi:10.1002/0471140864.ps0909s73
35. Gustafsson C, Govindarajan S, Minshull J (2004) Codon bias and heterologous protein expression. Trends Biotechnol 22(7):346–353. doi:10.1016/j.tibtech.2004.04.006
36. Miroux B, Walker JE (1996) Over-production of proteins in *Escherichia coli*: mutant hosts that allow synthesis of some membrane proteins and globular proteins at high levels. J Mol Biol 260(3):289–298. doi:10.1006/jmbi.1996.0399
37. Korpimaki T, Kurittu J, Karp M (2003) Surprisingly fast disappearance of beta-lactam selection pressure in cultivation as detected with novel biosensing approaches. J Microbiol Methods 53(1):37–42

Chapter 2

Expression of Plant Receptor Kinases in Yeast

María Laura Barberini and Jorge P. Muschietti

Abstract

The budding yeast *Saccharomyces cerevisiae* is a useful system to express recombinant proteins and analyze protein–protein interaction. Membrane-spanning proteins like plant receptor kinases find their way to the plasma membrane when expressed in yeast and seem to retain their structure and function. Here, we describe a general yeast DNA transformation procedure based on lithium acetate, salmon sperm DNA, and polyethylene glycol used to express recombinant proteins. Yeast cells expressing plant receptor kinases can be used for in vivo and in vitro studies of receptor function.

Key words *Saccharomyces cerevisiae*, Yeast, Transformation, Receptor kinase

1 Introduction

Yeast is an excellent eukaryotic model system widely used for molecular biology and genetics studies. Yeast was the first eukaryotic organism to have its genome sequenced. Yeast has a well-defined DNA transformation system that has been used for genetic manipulation. The first successful protocol for yeast genetic transformation was established using spheroplasts [1]. Several other transformation methods such as electroporation, biolistics, and agitation with glass beads have been developed [2].

Yeast is a suitable system to express plant proteins that require posttranslational modification as well as secreted or intracellular proteins. Furthermore, yeast provides other benefits such as cheap cost growth and maintenance and full tools for genetic modification approaches. In our laboratory, we employed yeast to successfully express proteins with one transmembrane domain (LePRK1 and LePRK2) [3, 4] and with six transmembrane domains such as NIP4;1 and NIP4;2, two *Arabidopsis thaliana* water channels [5]. LePRK1 and LePRK2 are two pollen-specific receptor-like kinases from *Solanum lycopersicum*, localized in the plasma membrane of pollen tubes and involved in pollen–pistil interactions [6–8]. The extracellular domains of LePRK1 and LePRK2 interact with

Reidunn Birgitta Aalen (ed.), *Plant Receptor Kinases: Methods and Protocols*, Methods in Molecular Biology, vol. 1621, DOI 10.1007/978-1-4939-7063-6_2, © Springer Science+Business Media LLC 2017

different pistil or pollen proteins along its path to the ovary [9–13]. In tomato pollen membranes, LePRK2 can be phosphorylated [3] exhibiting multiple phosphorylated isoforms [14]. The kinase activity of LePRK2 is necessary for interaction with LePRK1 [3, 4]. We also found that the juxtamembrane domain of LePRK1 was sufficient for its interaction with LePRK2 [4].

Here we describe a general yeast DNA transformation procedure based on lithium acetate (LiAC), salmon sperm DNA (ssDNA), and polyethylene glycol (PEG) [15]. To express protein receptor-like kinases in yeast, we use the inducible promoter pGAL1 (galactosidase promoter). In the presence of glucose, the pGAL1 promoter is repressed while transcription is highly induced when yeast is grown on galactose as the only carbon source.

2 Materials

2.1 Yeast Strain

The protease-deficient, galactose-inducible yeast strain BJ2168 (ATCC 208277) is used for transformation. The genotype of this strain is MATa *prc1407 prb11122 pep43 leu2 trp1 ura352 gal2* (*see* **Note 1**). Yeast strain is stored for long periods of time in YPAD medium with 20% glycerol at −70 °C. Transformed yeast strains are stored in the appropriate SD dropout medium with 20% glycerol at −70 °C.

2.2 Plasmids

For transformation, we use a series of four yeast cloning vectors (YCpIF vectors) each containing a different auxotropy marker [16]. The genes to be expressed are under control of the galactose-inducible yeast GAL1 promoter. Each of the YCpIF vectors derives from one of the series of pUN vectors (*Escherichia coli*-yeast shuttle vectors) [17]. They all contain both yeast and bacterial origins of replication (ARS1 and ori, respectively), a yeast centromere (CEN4), the bacterial AmpR gene, and four different yeast selectable markers (YSM: leucine, uracil, histidine, and tryptophan). They have a translation start codon upstream from a multiple cloning site (MCS) (*see* Fig. 1). To study, for example, the interaction between two receptors, their cDNA sequence is cloned in vectors with different markers and transformed into yeast (*see* **Note 2**).

2.3 Reagents

1. YPAD medium (per liter): 20 g peptone, 10 g yeast extract, and water to 960 mL; adjust pH to 5.8. Add 40 mg of adenine sulfate. Autoclave. Cool to 55 °C. Add sterilized glucose 50% stock solution to final 2% (v/v). Add water to 1 L. For YPAD agar add 2% of agar.
2. SD medium (per liter): 6.7 g of yeast nitrogen base without amino acids, 182.2 g of D-sorbitol, and water to 860 mL; adjust pH to 5.8. Autoclave. Cool to 55 °C. Add sterilized

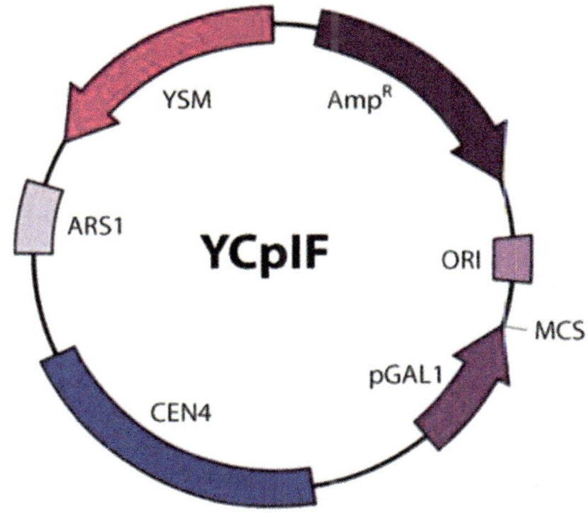

Fig. 1 Schematic representation of yeast cloning vectors (YCpIF). *pGAL1* yeast galactose-inducible GAL1 promoter, *MCS* multiple cloning site, *ori* bacterial origin of replication, *ARS1* a yeast autonomously replicating sequence, *CEN 4* a centromere derived from yeast, *YSM* yeast selectable markers, *Amp*^{R} ampicillin-resistance gene

glucose 50% (w/v) stock solution to final 2% (v/v) and 100 mL of the required 10× dropout solution. Add water to 1 L. For SD agar add 2% (w/v) of agar.

3. 10× dropout solution (DO): Use a commercial dropout solution that has all the nutrients with the exception of leucine, uracil, and tryptophan (Clontech, -Leu/-Trp/-Ura DO Supplement catalog #630426). Add tryptophan to 200 mg/L. The final DO solution is autoclaved and stored at 4 °C for up to 1 year.
4. Sterile LiAC: Prepare 1 M lithium acetate, pH 8.4–9.0. Autoclave.
5. Sterile PEG 50% (w/v): Dissolve 25 g of PEG MW 3350 in 18 mL of water. Mix with magnetic bar to completely dissolve it. Add water to 50 mL. Filter sterilize it.
6. ssDNA (2 mg/mL): Dissolve 20 mg of salmon sperm DNA in 10 mL of water. Mix with magnetic bar to completely dissolve it. Aliquot and store at −20 °C.
7. Yeast buffer: 20 mM Tris–HCl, pH 8.0, 10 mM $MgCl_2$, 1 mM EDTA, 5% (v/v) glycerol, 1 mM DTT, 300 mM $(NH_4)_2SO_4$, and 1× protease inhibitor mixture.
8. Yeast membrane buffer: 50 mM Tris–HCl, pH 8.0, and 150 mM NaCl, 0.5% (v/v) Nonidet P-40.
9. Tris-buffered saline (TBS): 1× 50 mM Tris–HCl, pH 7.5, and 150 mM NaCl.

10. Tris-buffered saline with Triton X-100 (TBST) buffer: 1× TBS with 0.2% (v/v) Triton X-100.
11. Blocking buffer: 6% (w/v) nonfat dry milk and 4% (w/v) glycine in TBST.
12. Antibody dilution buffer: 0.3% (w/v) nonfat dry milk and 0.3% (w/v) glycine in TBST.

3 Methods

3.1 Yeast Transformation

1. Start an overnight culture inoculating 2 mL of YPAD medium and incubating at 30 °C with shaking (*see* **Note 3**).
2. The following morning, inoculate 5 mL of YPAD medium to a cell density of 10^6 cells/mL culture, that is, an OD of 0.2–0.4 for the 208277 strain (*see* **Note 4**). Incubate the culture with shaking at 30 °C for approximately 4 h until 10^7 cells/mL.
3. Centrifuge for 10 min at 3000 × *g*.
4. Discard the growth media and wash the cells with 10 mL of sterile water. Centrifuge the cells again.
5. Discard the water and resuspend the cells in 0.1 mL of 100 mM LiAc. Transfer the cells to a sterile 1.5-mL microfuge tube.
6. Centrifuge the cells for 10 min at 5000 rpm. Remove the liquid.
7. Resuspend the cells to a final volume of 50 μL of 100 mM LiAc. The volume of LiAc must be adjusted to get a final cell concentration in the order of 10^9 cells/mL.
8. Centrifuge the cells for 10 min at 5000 rpm. Remove the LiAc.
9. Add the following reagents to the pellet in this order.
   - 240 μL PEG (50% w/v).
   - 36 μL 1 M LiAc.
   - 50 μL ssDNA (2 mg/mL) (*see* **Note 5**).
   - *x* μL plasmid DNA (1–10 μg).
   - 34–*x* μL sterile water.
10. Vortex vigorously the tube for 1 min or until the yeast pellet is dissolved.
11. Incubate the cell suspension for 30 min at 30 °C.
12. Heat shock the cells for 20 min at 42 °C.
13. Centrifuge the cells for 10 s at 5000 rpm. Remove the liquid.
14. Add 200 μL of sterile water and carefully resuspend the cells with a micropipette.

15. Plate different volumes of the cell suspension into the appropriate SD agar/glucose/dropout plates (*see* **Note 6**). For volumes smaller than 50 μL, add 100 μL of water to help plating.
16. Incubate the plates for 2–3 days at 30 °C to obtain transformants.

3.2 Expression of Receptor Kinases

1. Once transformed cells are obtained, take 2–4 colonies for each plate, and inoculate 5 mL of liquid SD/glucose/dropout medium. Incubate for 2 days at 30 °C with shaking.
2. Centrifuge the cells for 10 min at 5000 rpm, and resuspend in an SD/galactose/dropout medium, to induce gene expression of the corresponding genes under the GAL1 promoter. Grow for additional 24 h.
3. Centrifuge the cells for 10 min at 5000 rpm, and resuspend in two volumes of yeast buffer.
4. Add four volumes of chilled 0.5-mm glass beads (Sigma). Vortex vigorously the tube for 1 min at 4 °C. Discard supernatant.
5. Wash glass beads with one volume of yeast buffer, and pool both supernatants.
6. Centrifuge at 10,000 × *g* for 10 min at 4 °C, and fractionate the S_{10} fractions by ultracentrifugation at 100,000 × *g* for 3 h at 4 °C.
7. Resuspend the yeast microsomal membranes (P_{100}) in yeast membrane buffer by stirring on a magnetic stirrer at 4 °C for 1 h.
8. Adjust protein concentration to 5–10 mg/mL.

3.3 Immunoblot of Expressed Receptor Kinases

1. Protein (40–60 μg) of the yeast microsomal fractions are separated by SDS-PAGE and blotted to nitrocellulose.
2. The membranes are blocked in blocking buffer for 30 min.
3. The blocked membranes are incubated with proper antibodies (*see* **Note 7**) diluted to 1:1000 in antibody dilution buffer for 1 h with shaking at room temperature.
4. After six washes of 10 min each with TBST, the membranes are incubated for 1 h with sheep anti-mouse polyclonal secondary antibodies conjugated with horseradish peroxidase (Amersham Biosciences), washed, and developed by using the enhanced chemiluminescence kit (Amersham Biosciences).

4 Notes

1. The BJ2168 yeast strain carries the auxotrophic markers, leucine (*leu2*), uracil (*ura352*), and tryptophan (*trp1*), for selection of yeast, which have been transformed with the plasmids YCpIF3,

YCpIF6, and YCpIF12, respectively. It is also strain proteinase deficient (CH1), homozygous *prb1* that alters sporulation and homozygous *pep4* that prevents it.

2. To study the tomato pollen receptors, we cloned them into the vector with leucine marker (LePRK1) and the uracil marker (LePRK2).
3. These volumes are for one transformation.
4. Because OD varies among different strains, use a Neubauer chamber to count the yeast cells.
5. Salmon sperm DNA must be boiled 5 min and immediately kept on ice.
6. In our case, SD -Leu, -Ura, using a commercial -Leu/-Trp/-Ura DO Supplement (Clontech catalog #630426) to whom additional tryptophan was added.
7. We used our own antibodies against LePRK1 and LePRK2 [6].

Acknowledgments

This work was supported by grants to J.M. (UBACyT, PICT2012, PICT2014, and PICT2015).

References

1. Hinnen A, Hicks JB, Fink GR (1978) Transformation of yeast. Proc Natl Acad Sci U S A 75:1929–1933
2. Kawai S, Hashimoto W, Murata K (2010) Transformation of Saccharomyces cerevisiae and other fungi: methods and possible underlying mechanism. Bioeng Bugs 1(6):395–403
3. Wengier D, Valsecchi I, Cabanas ML, Tang WH, McCormick S, Muschietti J (2003) The receptor kinases LePRK1 and LePRK2 associate in pollen and when expressed in yeast, but dissociate in the presence of style extract. Proc Natl Acad Sci U S A 100(11):6860–6865
4. Salem TM, Barberini ML, Wengier DL, Cabanas ML, de Paz P, Muschietti J (2012) Oligomerization studies show that the kinase domain of the tomato pollen receptor kinase LePRK2 is necessary for interaction with LePRK1. Plant Physiol Biochem 53:40–45
5. Di Giorgio JA, Bienert GP, Ayub ND, Yaneff A, Barberini ML, Mecchia MA, Amodeo G, Soto GC, Muschietti JP (2016) Pollen-specific aquaporins NIP4;1 and NIP4;2 are required for pollen development and pollination in Arabidopsis thaliana. Plant Cell 28(5):1053–1077
6. Muschietti J, Eyal Y, McCormick S (1998) Pollen tube localization implies a role in pollen-pistil interactions for the tomato receptor-like protein kinases LePRK1 and LePRK2. Plant Cell 10(3):319–330
7. Zhang D, Wengier D, Shuai B, Gui CP, Muschietti J, McCormick S, Tang WH (2008) The pollen receptor kinase LePRK2 mediates growth-promoting signals and positively regulates pollen germination and tube growth. Plant Physiol 148(3):1368–1379
8. Gui CP, Dong X, Liu HK, Huang WJ, Zhang D, Wang SJ, Barberini ML, Gao XY, Muschietti J, McCormick S, Tang WH (2014) Overexpression of the tomato pollen receptor kinase LePRK1 rewires pollen tube growth to a blebbing mode. Plant Cell 26(9):3538–3555
9. Tang W, Ezcurra I, Muschietti J, McCormick S (2002) A cysteine-rich extracellular protein, LAT52, interacts with the extracellular domain of the pollen receptor kinase LePRK2. Plant Cell 14(9):2277–2287
10. Tang W, Kelley D, Ezcurra I, Cotter R, McCormick S (2004) LeSTIG1, an extracellular

binding partner for the pollen receptor kinases LePRK1 and LePRK2, promotes pollen tube growth in vitro. Plant J 39(3):343–353

11. Wengier DL, Mazzella MA, Salem TM, McCormick S, Muschietti JP (2010) STIL, a peculiar molecule from styles, specifically dephosphorylates the pollen receptor kinase LePRK2 and stimulates pollen tube growth in vitro. BMC Plant Biol 10:33
12. Huang WJ, Liu HK, McCormick S, Tang WH (2014) Tomato pistil factor stig1 promotes in vivo pollen tube growth by binding to phosphatidylinositol 3-phosphate and the extracellular domain of the pollen receptor kinase LePRK2. Plant Cell 26(6): 2505–2523
13. Zhao XY, Wang Q, Li S, Ge FR, Zhou LZ, McCormick S, Zhang Y (2013) The juxtamembrane and carboxyterminal domains of Arabidopsis PRK2 are critical for ROP-induced growth in pollen tubes. J Exp Bot 64(18):5599–5610
14. Salem T, Mazzella A, Barberini ML, Wengier D, Motillo V, Parisi G, Muschietti J (2011) Mutations in two putative phosphorylation motifs in the tomato pollen receptor kinase LePRK2 show antagonistic effects on pollen tube length. J Biol Chem 286(6):4882–4891
15. Gietz RD, Schiestl RH, Willems AR, Woods RA (1995) Studies on the transformation of intact yeast cells by the LiAc/SS-DNA/PEG procedure. Yeast 11:355–360
16. Foreman PK, Davis RW (1994) Cloning vectors for the synthesis of epitope-tagged, truncated and chimeric proteins in Saccharomyces cerevisiae. Gene 144(1):63–68
17. Elledge SJ, Davis RW (1988) A family of versatile centromeric vectors designed for use in the sectoring-shuffle mutagenesis assay in Saccharomyces cerevisiae. Gene 70(2):303–312

Chapter 3

Expression of Plant Receptor Kinases in Tobacco BY-2 Cells

Hidefumi Shinohara and Yoshikatsu Matsubayashi

Abstract

Although more than 600 single-transmembrane receptor kinase genes have been found in the *Arabidopsis* genome, only a few of them have known physiological functions, and even fewer plant receptor kinases have known specific ligands. Ligand-binding analysis must be operated using the functionally expressed receptor form. However, the relative abundance of native receptor kinase molecules in the plasma membrane is often quite low. Here, we present a method for stable and functional expression of plant receptor kinases in tobacco BY-2 cells that allows preparation of microsomal fractions containing the receptor. This procedure provides a sufficient amount of receptor proteins while maintaining its ligand-binding activities.

Key words Receptor kinase, Peptide ligand, Ligand binding, HaloTag

1 Introduction

Cell-to-cell communication is essential for the growth and development of multicellular organisms throughout their lives. Extracellular signals are usually perceived by membrane-localized receptors that transduce the signals inside cells to activate programs that direct growth and development. More than 600 single-transmembrane receptor kinase (RK) genes have been found in the *Arabidopsis* genome [1]. Despite the large number of RKs in plants, fewer than 50 of them have known physiological functions, and even fewer plant RKs have known specific ligands.

As ligand recognition by the membrane receptor is highly dependent on the correctly folded protein surface, molecular interactions must be analyzed using the functionally expressed receptor form, preferably with the native receptor protein. However, the relative abundance of specific RK molecules in the plasma membrane is often quite low, even in transgenic plants that overexpress target RKs.

Reidunn Birgitta Aalen (ed.), *Plant Receptor Kinases: Methods and Protocols*, Methods in Molecular Biology, vol. 1621, DOI 10.1007/978-1-4939-7063-6_3,

Here, we present a method for stable and functional expression of plant RKs in tobacco BY-2 cells. We introduce (1) how to express plant tag-fused RKs in tobacco BY-2 cells, (2) how to prepare the membrane fraction containing RKs, and (3) how to detect RKs. Tag-fused plant RKs in which the kinase domain is replaced with an affinity tag are expressed in BY-2 cells while maintaining its ligand-binding activities. Tag-fused plant RKs expressed in BY-2 cells also provide a sufficient amount of receptor protein for biochemical analysis. Previous works showed that ligand-binding experiments succeeded using tag-fused RKs [2–8] (*see* also Chapter 8, Photoaffinity Labeling of Plant Receptor Kinases by ^{125}I–labeled Photoreactive 4-azidosalicylic acid ([^{125}I]ASA)-marked Peptide Ligands).

2 Materials

2.1 Transformation of BY-2 Cells

1. Tobacco BY-2 cells (*see* **Note 1**).
2. Murashige and Skoog (MS) salt (Wako 392–00591).
3. KH_2PO_4 stock: 200 mg/ml (dissolve in water).
4. Myo-inositol stock: 20 mg/ml (dissolve in water).
5. Thiamine-HCl stock: 1 mg/ml (dissolve in water).
6. 2,4-Dichlorophenoxyacetic acid (2,4-D) stock: 0.2 mg/ml (dissolve in water).
7. Sucrose.
8. Sterile dish.
9. Modified MS (mMS) medium: Add 1 bag MS salt, 1 ml of KH_2PO_4 stock, 5 ml of myo-inositol stock, 1 ml of Thiamine-HCl stock, 1 ml of 2,4-D stock, and 30 g of sucrose to a 1-l glass beaker. Add water to a volume of 900 ml. Mix and adjust pH to 5.8 with aqueous KOH. Make up to 1 l with water and then autoclave.
10. mMS plates: 0.4% gellan gum with appropriate antibiotics.
11. *Agrobacterium* strain C58C1 transformed with a binary vector containing the HaloTag-fused RK construct under the control of the CaMW 35S promoter (colonies on LB agar plates) (*see* **Notes 2–4**).
12. Luria-Bertani (LB) broth: 2% tryptone, 1% yeast extract, and 2% NaCl.
13. 1000× antibiotics: Dissolve in water or appropriate solvent and sterilize using a 0.22-μm syringe filter.
14. Incubator set at 25 °C.
15. Incubator shaker set at 25 °C.

2.2 Preparation of Microsomal Fraction from Receptor-Expressed BY-2 Cells

1. Sterilized mMS medium.
2. 100-ml flask.
3. 300-ml flask.
4. Filter paper (ADVANTEC No. 2).
5. Extraction buffer: 25 mM Tris-HCl, pH 7.0, 10 mM $MgCl_2$, 2 mM dithiothreitol, 2 μM leupeptin, 2 mM phenylmethanesulfonyl fluoride, and 250 mm sucrose (*see* **Note 5**).
6. Homogenizer (NISSEI, AM-3).
7. Miracloth (Calbiochem, 23333–11).
8. Centrifugation tubes (Beckman 357001).
9. Ultracentrifuge (Beckman Avanti HP-30).
10. Suspension buffer: 10 mM Tris–HCl, pH 7.0, 250 mM sucrose (*see* **Note 6**).
11. Protein Assay Kit (Bio-Rad 5000001JA).

2.3 Visualization of HaloTag-Fused Receptor by Incorporating the Fluorescent Dye TMR

1. TMR reaction buffer: 20 mM HEPES-KOH, pH 7.0.
2. HaloTag TMR ligand: 50 μM stock in DMSO (Promega, G8251).
3. Incubator set at 37 °C.
4. 2× SDS sample buffer: 0.1 M Tris-HCl, pH 6.8, 4% SDS, 12% ß-mercaptoethanol, 8% glycerol, and 0.02% bromophenol blue.
5. Heat block set at 95 °C.
6. 7.5% acrylamide gel (*see* **Note 7**).
7. SDS-PAGE running buffer: 25 m Tris, pH 8.2, 0.192 M glycine, and 0.1% SDS.
8. Imaging analyzer (GE, Typhoon 9400).

3 Methods

3.1 Transformation of BY-2 Cells

1. BY-2 cells are maintained in a 1-week culture interval under continuous darkness at 22 °C with shaking at 130 rpm. Agitate a 1-week-old culture well and transfer 1.2 ml of BY-2 cell suspension to 95 ml of fresh mMS medium in a 300-ml flask with a pipette.
2. Streak *Agrobacterium* carrying binary vector containing the HaloTag-fused RK construct onto LB medium containing appropriate antibiotics for the plasmid of interest plus 50 μg/ml gentamicin and 50 μg/ml rifampicin (*see* **Note 8**).
3. Pick colonies from the plate and transfer into 2 ml of LB medium containing appropriate antibiotics for plasmids. Culture for 1 day at 30 °C.

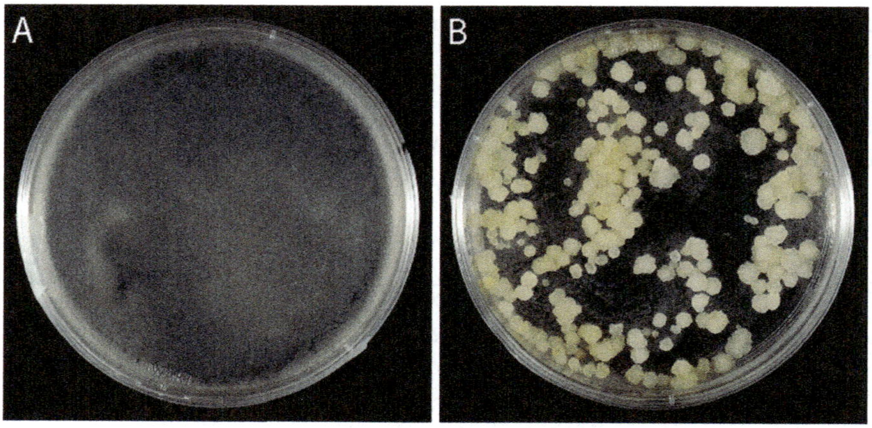

Fig. 1 Selection of transformed BY-2 cells. *Agrobacterium*-treated BY-2 cells spread on the plate after 5 days (**a**) and after 5 weeks (**b**)

4. Pipette 6 ml of BY-2 cells into a sterile dish 3 days after inoculation.
5. Add 200 μl of saturated *Agrobacterium* culture into BY-2 cells.
6. Incubate at 25 °C for 2 days in the dark without shaking.
7. Transfer the cocultivated cells into 15-ml tubes, and centrifuge at 150 × *g* for 3 min.
8. Remove the supernatant and add 10 ml of mMS medium.
9. Mix gently by inversion.
10. Centrifuge at 150 × *g* for 3 min.
11. Repeat **steps 8–10** three times.
12. Add mMS medium up to 5 ml.
13. Spread 2 and 3 ml of BY-2 cells on an MS plate containing appropriate antibiotics plus 500 μg/ml carbenicillin (*see* **Note 9**).
14. Incubate at 25 °C in the dark. Transformed calli will be visible after 3 to 4 weeks (*see* Fig. 1).

3.2 Preparation of Microsomal Fraction from Receptor-Expressing BY-2 Cells

1. Transfer transformed calli to 50 ml of fresh mMS medium in a 100-ml flask.
2. Incubate at 25 °C for 7 days in the dark with shaking at 130 rpm.
3. Transfer 1.2 ml of transformed BY-2 cell suspension to 95 ml of fresh mMS medium in a 300-ml flask.
4. Incubate at 25 °C for 7 days in the dark with shaking at 130 rpm.
5. Remove the medium and collect BY-2 cells with filter paper.
6. Cool the collected cells on ice within a homogenizer bucket.

7. Add 20 ml of ice-cold extraction buffer. Mix and cool the suspension well on ice.
8. Homogenize with a homogenizer at 4 °C.
9. Pass the homogenate through Miracloth.
10. Pour the filtrate into centrifugation tubes and centrifuge at 5000 × *g* for 15 min at 4 °C to precipitate cell debris and organelles, such as nuclei, mitochondria, and chloroplasts.
11. Collect the supernatant and transfer to new centrifugation tubes.
12. Centrifuge at 100,000 × *g* for 30 min at 4 °C to precipitate the microsomal fraction.
13. Remove the supernatant.
14. Suspend the pellet containing the microsomal fraction in 1.0 ml of suspension buffer by gentle pipetting.
15. Determine the protein concentration in a conventional Bradford assay according to the manufacturer's protocol (*see* **Note 10**).
16. Store at −80 °C.

3.3 Visualization of HaloTag-Fused Receptor by Incorporating the Fluorescent Dye TMR

1. Suspend the aliquots of microsomal fractions from receptor-expressing BY-2 cells (100 μg eq.) in 20 μl of TMR reaction buffer.
2. Add 2 μl of 50 μM HaloTag TMR ligand to a final concentration of 5 μM.
3. Mix and incubate at 37 °C for 30 min.
4. Add 20 μl of 2× SDS sample buffer and mix.
5. Heat at 95 °C for 5 min.
6. Load the heated sample on a 7.5% acrylamide gel and separate by SDS-PAGE.
7. Analyze the gel using a Typhoon 9400 (Amersham) with a 523-nm excitation filter and a 580-nm emission filter (*see* Fig. 2) (*see* **Note 11**).

4 Notes

1. Tobacco BY-2 cells can be obtained from RIKEN BRC (http://epd.brc.riken.jp/en/by_2).
2. In the preparation of the HaloTag-fused RK construct, the cytoplasmic kinase domain of RK should be replaced with the HaloTag. The kinase domain of RK is not required for ligand recognition, which is often the case with single-transmembrane-spanning receptors. In addition, deletion of the cytoplasmic

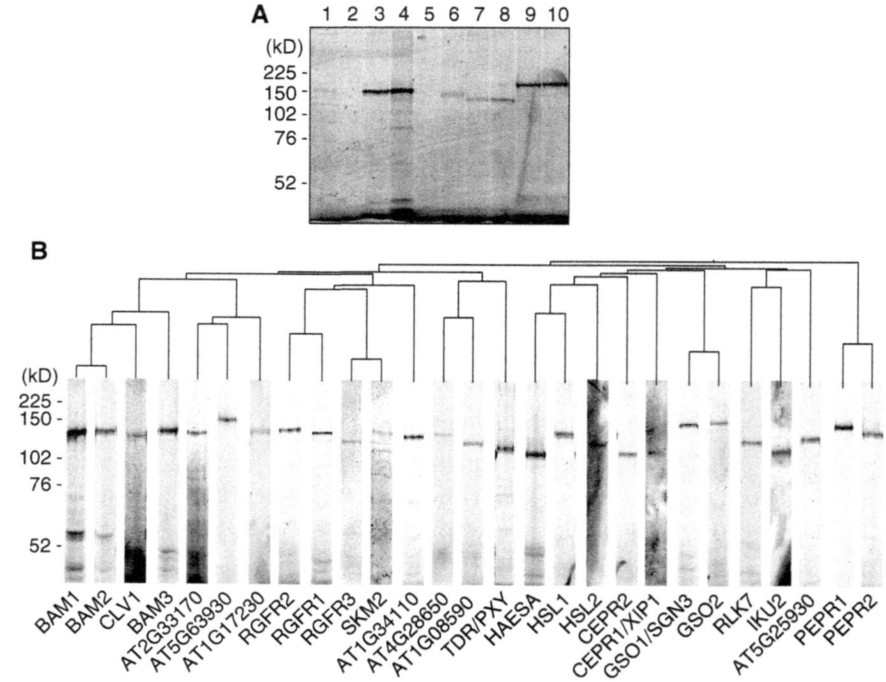

Fig. 2 Visualization of HaloTag-fused receptors in transgenic tobacco BY-2 cells. (**a**). Aliquots of microsomal proteins (100 μg eq.) were incubated with HaloTag TMR ligand and separated by SDS-PAGE. Samples in each lane are derived from individual HaloTag-fused receptor expression lines. (**b**). HaloTag-fused RK expression library covering 27 *Arabidopsis* leucine-rich repeat receptor kinases (LRR-RKs) in subgroup XI

kinase domain improves expression of RK proteins, likely because of the absence of the kinase domain, which may be involved in the regulation of RK turnover [2].

3. The HaloTag, a 33-kDa protein derived from haloalkane dehalogenase, can form a covalent bond with a chloroalkane group via the substitution of chlorine catalyzed by intrinsic enzyme activity [9].
4. Other affinity tags, for example, His-tag [10], are also possible to use for this experiment.
5. Extraction buffer should be prepared freshly for each experiment.
6. Once prepared, the suspension buffer can be stored in −20 °C.
7. A 7.5% acrylamide gel is suitable for the separation of HaloTag-fused receptor kinases. The concentration of acrylamide in an SDS-PAGE gel is dependent on the molecular weights of the target proteins.
8. Gentamicin and rifampicin are selective antibiotics for the *Agrobacterium* strain C58C1 (pMP90). Other *Agrobacterium*

strains, e.g., GV3101, EHA105, or LBA4404, are suitable for BY-2 cell transformation.

9. The antibiotic carbenicillin can prevent *Agrobacterium* strain C58C1 (pMP90) growth.
10. Measure the protein concentration using the Bradford method [11], using bovine serum albumin (BSA) as a reference. We recommend that the protein concentration of the microsomal fraction be adjusted to ≈50 μg/μl.
11. The advantage of using HaloTag is that we can check the protein expression without immunoblotting. We have confirmed that visualization using the HaloTag TMR ligand shows the same sensitivity as that of immunoblot analysis using Anti-HaloTag antibodies.

Acknowledgments

This research was supported by a Grant-in-Aid for Scientific Research (S) (25221105 to Y.M.), a Grant-in-Aid for Scientific Research on Innovative Areas (15H05957 to Y.M., 26113520 and 16H01234 to H.S.), and a Grant-in-Aid for Young Scientists (B) (25840111 to H.S.).

Reference

1. Shiu S, Bleecker A (2001) Receptor-like kinases from Arabidopsis form a monophyletic gene family related to animal receptor kinases. Proc Natl Acad Sci U S A 98:10763–10768
2. Shinohara H, Matsubayashi Y (2007) Functional immobilization of plant receptor-like kinase onto microbeads towards receptor array construction and receptor-based ligand fishing. Plant J 52:175–184
3. Hirakawa Y, Shinohara H, Kondo Y, Inoue A, Nakanomyo I, Ogawa M, Sawa S, Ohashi-Ito K, Matsubayashi Y, Fukuda H (2008) Non-cell-autonomous control of vascular stem cell fate by a CLE peptide/receptor system. Proc Natl Acad Sci U S A 105:15208–15213
4. Ogawa M, Shinohara H, Sakagami Y, Matsubayashi Y (2008) Arabidopsis CLV3 peptide directly binds CLV1 ectodomain. Science 319:294
5. Shinohara H, Moriyama Y, Ohyama K, Matsubayashi Y (2012) Biochemical mapping of a ligand-binding domain within Arabidopsis BAM1 reveals diversified ligand recognition mechanisms of plant LRR-RKs. Plant J 70:845–854
6. Tabata R, Sumida K, Yoshii T, Ohyama K, Shinohara H, Matsubayashi Y (2014) Perception of root-derived peptides by shoot LRR-RKs mediates systemic N-demand signaling. Science 346:343–346
7. Shinohara H, Matsubayashi Y (2015) Reevaluation of the CLV3-receptor interaction in the shoot apical meristem: dissection of the CLV3 signaling pathway from a direct ligand-binding point of view. Plant J 82:328–336
8. Shinohara H, Mori A, Yasue N, Sumida K, Matsubayashi Y (2016) Identification of three LRR-RKs involved in perception of root meristem growth factor in Arabidopsis. Proc Natl Acad Sci U S A 113:3897–3902
9. Los G, Darzins A, Karassina N et al (2005) HaloTagTM inter-changeable labeling technology for cell imaging and protein capture. Promega Cell Notes 11:2–6
10. Shinohara H, Ogawa M, Sakagami Y, Matsubayashi Y (2007) Identification of ligand binding site of phytosulfokine receptor by on-column photoaffinity labeling. J Biol Chem 282:124–131
11. Bradford MM (1976) A rapid and sensitive method for the quantitation of microgram quantities of protein utilizing the principle of protein-dye binding. Anal Biochem 72:248–254

Chapter 4

Cell-Free Synthesis of Plant Receptor Kinases

Akira Nozawa, Keiichirou Nemoto, Shunsuke Nomura, Satoshi Yamanaka, Kohki Kido, and Tatsuya Sawasaki

Abstract

The wheat germ cell-free protein synthesis system has been used as a eukaryotic protein production system since it was first reported in 1964. Although initially the productivity of this system was not very high, it has now become one of the most versatile protein production systems, thanks to the enhancements made by several groups. In this chapter, we report a protein production method for plant receptor kinases using the wheat cell-free system. We describe a method for the preparation of a cell-free extract from wheat germ, the split-primer PCR method for preparation of transcription templates, and the bilayer cell-free protein synthesis method.

Key words Cell-free protein synthesis system, Wheat germ extract

1 Introduction

Protein synthesis using a wheat germ extract was first reported by Marcus and Feely in 1964 [1]. However, because the initial protein production achieved using this wheat germ cell-free system was very low, the use of radioactive isotopes had been essential for detection of the synthesized proteins. In order to increase productivity, the wheat germ cell-free system was enhanced by subsequent modifications made by several groups. In 1988, Spirin et al. [2] reported a continuous flow cell-free protein synthesis method. In this method, the reaction time was prolonged, and the yield of synthesized protein was increased by continuous supply of substrates and removal of small by-products using an ultrafiltration membrane. Kawarasaki et al. [3] demonstrated that the translation activity of wheat germ cell-free system could be increased by optimizing the concentrations of creatine phosphate and tRNAs and by attaching the 5′-untranslated region of tobacco etch virus to uncapped mRNAs. Nakano et al. [4] increased the productivity of the wheat germ cell-free system by using highly condensed wheat germ extract and copper ion. Addition of copper ion to the cell-free

Reidunn Birgitta Aalen (ed.), *Plant Receptor Kinases: Methods and Protocols*, Methods in Molecular Biology, vol. 1621, DOI 10.1007/978-1-4939-7063-6_4, © Springer Science+Business Media LLC 2017

system prevented the degradation of mRNA and ATP by reducing the activities of ribonucleases and phosphatases. In our laboratory, we succeeded in improving the wheat germ cell-free system by eliminating translational inhibitors such as ribosome-inactivating proteins and ribonucleases originating from the endosperm [5]. In this system, the translation reaction proceeded for 60 h, and the yield of the synthesized protein was greater than 1 mg. In addition, we obtained an artificial E02 sequence that stimulated translation initiation of uncapped mRNAs when attached to the 5′-untranslated region [6]. Furthermore, we developed a new strategy to construct templates for transcription using the split-primer PCR method [7] and established a simplified version of the continuous flow cell-free protein synthesis method, the bilayer cell-free synthesis method, without the need for dialysis [8]. With these improvements, the wheat germ cell-free system can today be used as a protein preparation method for structural analysis and high-throughput proteomics [9, 10]. Moreover, cell-free systems have certain advantages over in vivo expression systems in recombinant organisms. As cell-free systems are decoupled from host cell physiology, these systems are available for the production of a wide variety of proteins, with the exception of those proteins that have inhibitory effects against translation machinery. In addition, it is easy to modify the reaction conditions of the cell-free systems by supplementation of chaperons, detergents, and lipids. Incorporation of unnatural amino acids into the synthesized proteins is also possible. The wheat germ cell-free system has accordingly now become one of the most versatile protein production methods.

In plants, protein kinases regulate a wide variety of physiological and developmental processes [11, 12]. For the fine-tuning of these processes, plants utilize a large number of protein kinases. Genes encoding protein kinases are one of the largest gene families in plants. For example, the genomes of *Arabidopsis thaliana* and *Oryza sativa* encode over 1000 and 1400 members of the protein kinases, respectively [13, 14]. Although there are a large number of different classes of protein kinases, the largest class is the receptor kinases. *A. thaliana* and *O. sativa* possess more than 600 and 1000 members of receptor kinases, respectively [14, 15]. Receptor kinases have a ligand-binding and/or signal-accepting extracellular domain, a single transmembrane domain, and an intracellular kinase domain. Ligand binding to the extracellular domain induces homo- or heterodimerization and stimulates kinase activity of the intracellular domain. Through the kinase activity, receptor kinases transmit information from the extracellular environment to the cell interior. To date, genetic analysis has revealed that receptor kinases in plants participate in a variety of biological processes such as hormone signaling, symbiosis, disease resistance, self-incompatibility, and regulation of cell growth. However, little is known about the biochemical and structural characters of receptor kinases.

To analyze the biochemical function and structure of a protein, we need a sufficient amount of the protein in a functional state. In a previous study, we reported cell-free synthesis of 759 *Arabidopsis* protein kinases and detected the autophosphorylation activity of 15 receptor kinases [16]. These results indicate that our cell-free system is a promising tool for the production of a wide range of functional plant receptor kinases. In this chapter, we describe a method for the preparation of a cell-free extract from wheat germ, split-primer PCR for preparation of transcription templates, and the bilayer protein synthesis method. We anticipate that the wheat germ cell-free system can facilitate the characterization of plant receptor kinases.

2 Materials

2.1 Purification of Wheat Embryo Particles

1. Wheat seeds.
2. Rotor speed mill.
3. Vibratory sieve shaker.
4. Cyclohexane–carbon tetrachloride mixture (1:2.5, v/v).
5. 0.5% Nonidet P-40 (NP-40).
6. Sonicator.

2.2 Preparation of the Cell-Free Extract

1. Liquid nitrogen.
2. A mortar and pestle.
3. Extraction buffer (EB): 40 mM HEPES-KOH (pH 7.6), 100 mM potassium acetate, 5 mM magnesium acetate, 2 mM calcium chloride, 4 mM DL-dithiothreitol (DTT), and 0.3 mM amino acids.
4. Gel filtration buffer (GB): 30 mM HEPES-KOH (pH 7.6), 100 mM potassium acetate, 2.7 mM magnesium acetate, 0.4 mM spermidine, 2.5 mM DTT, 0.3 mM amino acids, 1.2 mM ATP, 0.25 mM GTP, and 16 mM creatine phosphate.
5. Sephadex G-25 columns (GE Healthcare).
6. Amicon Ultra-15 Centrifugal Filter Unit (Merck Millipore).

2.3 Preparation of Template for Transcription

1. The cDNA of the receptor of interest cloned in a plasmid designed for wheat germ cell-free protein synthesis (*see* **Note 1** and Fig. 1).
2. Primer set for first PCR. Sense primer [5′-CCACCCACC ACCACCAATGN~N-3′; the sequence ATGN~N indicates the 5′-terminal coding region of a target gene (generally 18 bp in length, including the initial ATG)] and antisense primer (design first antisense primer, which binds to a site on the plasmid vector 1–2 kb downstream from the 3′ end of a target gene) (*see* Fig. 1).

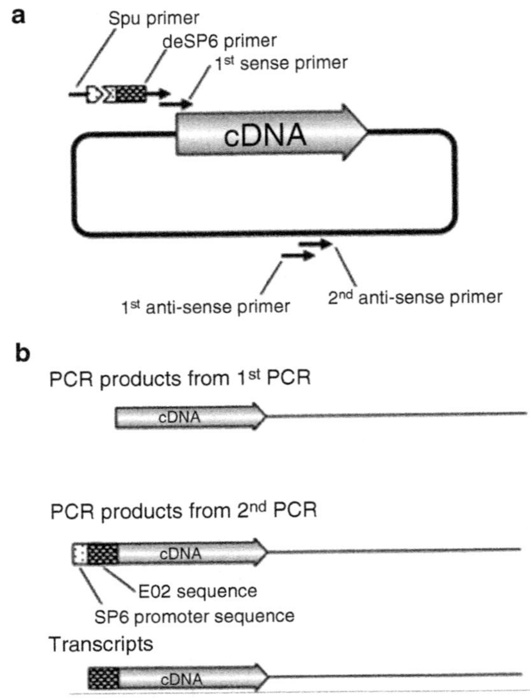

Fig. 1 Preparation of transcriptional templates using the split-primer PCR method. (**a**) Design of the split-type primers used for introducing the required UTRs into the cDNA sequences. (**b**) Expected PCR products and transcripts

3. Blend Taq (2.5 units/μL) and 10× Blend Taq buffer (Toyobo, Osaka, Japan).
4. 2 mM dNTP mixture: 2 mM each of dATP, dCTP, dGTP, and dTTP.
5. Primer set for second PCR. Spu primer (5′-GCGTAGCATTTAGGTGACACT-3′), deSP6 primer (5′-GGTGACACTATAGAACTCACCTATCTCTCTACACAAAACATTTCCCTACATACAACTTTCAACTTCCTATTCCACCCACCACCACCAATG-3′), and second antisense primer (design second antisense primer, which binds to an internal region 3 ~ 100 bp from the first antisense primer binding site) (*see* **Notes 1** and **2**).

2.4 Preparation of mRNA

1. 5× transcription buffer: 400 mM HEPES-KOH (pH 7.8), 80 mM magnesium acetate, 10 mM spermidine, and 50 mM DTT.
2. 25 mM NTP mixture: 25 mM each of ATP, CTP, GTP, and UTP.
3. Ribonuclease (RNase) inhibitor (80 units/μL) and SP6 RNA polymerase (80 units/μL) (Promega, Madison, WI).
4. 7.5 M ammonium acetate.

2.5 Protein Synthesis

1. Creatine kinase (Roche Diagnostics, Kyoto, Japan) is dissolved in water at 20 mg/mL and stored in aliquots at −80 °C.
2. 4× dialysis buffer (DB): 120 mM HEPES-KOH (pH 7.8), 400 mM potassium acetate, 10.8 mM magnesium acetate, 1.6 mM spermidine, 10 mM DTT, 4.8 mM ATP, 1 mM GTP, 64 mM creatine phosphate, 0.02% NaN_3, and 1.2 mM of each amino acid. Store at −80 °C.

3 Methods

3.1 Purification of Wheat Embryo Particles

1. Grind the wheat seeds (typically 6 kg for 6 mL of extract) by milling.
2. Sieve the sample using a sieve shaker, and collect the particles that pass through the 850-μm mesh but not the 710-μm mesh.
3. Remove lightweight flakes of the bran by dropping the sample from the height of a few feet into another container.
4. Isolate the intact embryos by solvent flotation using a mixture of cyclohexane and carbon tetrachloride (1:2.5, v/v) as a solvent.
5. Collect the floating particles and dry them overnight in a fume hood.
6. Pick up yellow embryo particles using toothpicks by visual judgment (Fig. 2): remove carefully those particles with a brownish color and those with considerable white matter.
7. Wrap the embryo particles (typically 2 g) in gauze.
8. Dip the embryo particles into Milli-Q water (typically 100 mL), and knead them gently but thoroughly to remove the white matter from the embryos.

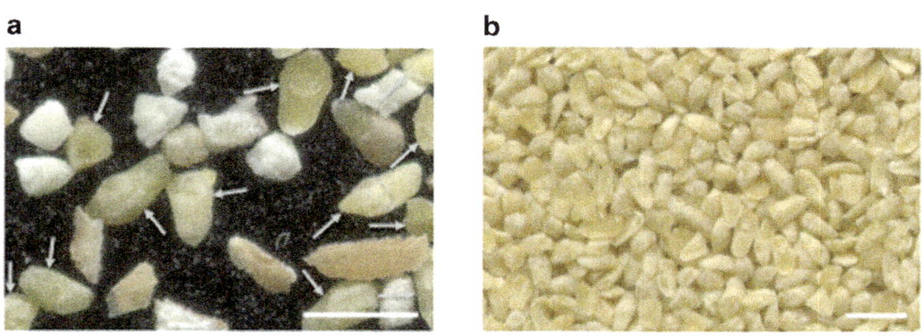

Fig. 2 Embryo particles. (**a**) Embryo particles before selection. Yellow embryo particles (indicated by *arrows*) should be used for preparation of wheat germ extracts. White and brownish granules are contaminants from endosperms and contain translational inhibitors such as ribosome-inactivating proteins and ribonucleases. (**b**) Embryo particles after selection. White and brownish granules are eliminated. There are only yellow embryo particles. Scale bar indicates 2.5 mm

9. Change the water and repeat washing twice more.
10. Sonicate the sample in 0.5% NP-40 using a sonicator for 2–3 min.
11. Repeat sonication twice more in Milli-Q water.
12. Put the embryo particles into a Buchner funnel, and wash them thoroughly with Milli-Q water (500 mL).

3.2 Preparation of the Cell-Free Extract

1. Grind the purified embryo particles (2 g) to a fine powder in liquid nitrogen using a precooled mortar and pestle.
2. Add 2 mL of chilled EB and briefly vortex the mixture.
3. Centrifuge the embryo lysate at 2000 × *g* for 1 h at 4 °C to remove lipids and recover the supernatant.
4. Centrifuge the sample at 30,000 × *g* for 30 min at 4 °C, and recover the supernatant.
5. Gel-filtrate the supernatant by using a Sephadex G-25 column pre-equilibrated with EB.
6. Collect the flow through fraction and repeat gel filtration with GB.
7. Centrifuge at 30,000 × *g* for 10 min and recover the supernatant.
8. Measure ultraviolet absorption of the supernatant at 260 nm, and adjust it to 240 with GB. If needed, concentrate the supernatant with an ultrafiltration module (Amicon Ultra-15) before adding GB.
9. Divide into small aliquots and store at −80 °C until use.

3.3 Preparation of Template for Transcription

1. Prepare 50 μL of the first PCR mixture as follows: 5 μL of 10× Blend Taq buffer, 5 μL of 2 mM dNTPs, 2.5 μL of 2 μM sense primer, 2.5 μL of 2 μM antisense primer, 1 μL of a template plasmid (1 ng/μL), 0.5 μL of Blend Taq (2.5 units/μL), and 33.5 μL of Milli-Q water.
2. Place the first PCR mixture into a PCR thermal cycler for a 2-min denaturation at 94 °C, followed by 30 cycles of amplification steps: 98 °C for 15 s, 55 °C for 30 s, and 72 °C for x min (1 min/kb target fragment).
3. Prepare 50 μL of the second PCR mixture as follows: 5 μL of 10× Blend Taq buffer, 5 μL of 2 mM dNTPs, 1 μL of 5 μM Spu primer, 1 μL of 2 μM second antisense primer, 0.5 μL of 0.1 μM deSP6 primer, 1 μL of first PCR product, 0.5 μL of Blend Taq (2.5 units/μL), and 36 μL of Milli-Q water (*see* **Note 3**).
4. Place the second PCR mixture into a PCR thermal cycler for a 2-min denaturation at 94 °C followed by 5 cycles of the first amplification steps and 30 cycles of the second amplification steps: 98 °C for 15 s, 55 °C for 1 min, and 72 °C for x min (1 min/kb target fragment) for the first amplification steps; 98 °C for 15 s, 60 °C for 30 s, and 72 °C for x min (1 min/kb target fragment) for the second amplification steps.

3.4 Preparation of mRNA

1. Prepare 25 μL of transcription mixture as follows: 5 μL of 5× transcription buffer, 2.5 μL of 25 mM NTP mix, 2.5 μL of the second PCR products, 0.25 μL of RNase inhibitor, 0.25 μL of SP6 RNA polymerase, and 14.5 μL of Milli-Q water.
2. Incubate at 37 °C for 3–6 h.
3. Add 4.7 μL of Milli-Q water, 5.3 μL of 7.5 M ammonium acetate, and 100 μL of ethanol to the transcription mixture, and place this on ice for 20 min.
4. Centrifuge the mixture at 20,000 × *g* for 20 min at 4 °C.
5. Remove the supernatant and dry the pellet for 1 min at room temperature.
6. Resuspend the pellet with 10 μL of Milli-Q water.

3.5 Cell-Free Protein Synthesis

1. Thaw wheat germ extract, creatine kinase, and dialysis buffer on ice. Keep all reagents used on ice. Freeze promptly after use.
2. Prepare 25 μL of translation mixture on ice as follows: 6.25 μL of wheat germ extract, 0.25 μL of 20 mg/mL creatine kinase, 4.7 μL of 4× DB, 10 μL of mRNA, and 3.8 μL of Milli-Q water.
3. Prepare 125 μL of substrate mixture on ice as follows: 31.25 μL of 4× DB and 93.75 μL of Milli-Q water.
4. Transfer the 125 μL of substrate mixture to a microtiter plate.
5. Carefully pipette the 25 μL of translation mixture underneath the substrate mixture to form a bilayer (Fig. 3).
6. Seal the plate to avoid evaporation.

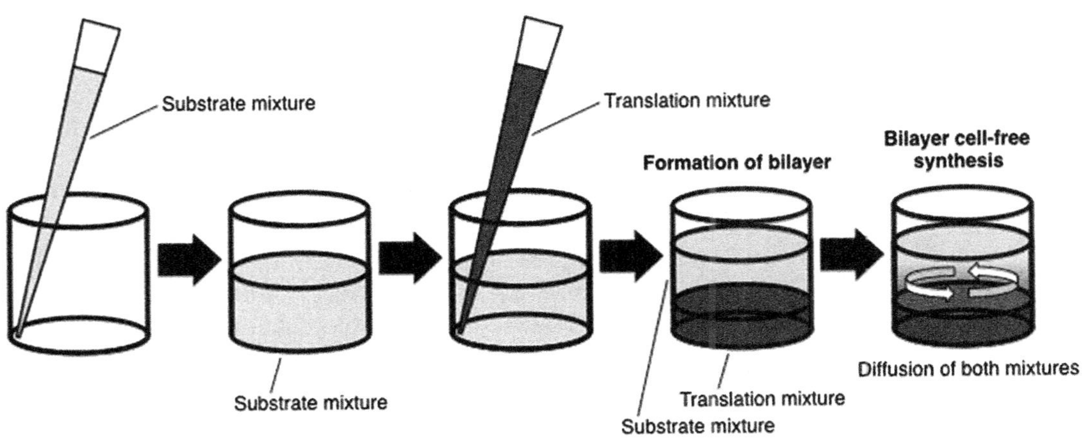

Fig. 3 Bilayer cell-free protein synthesis method. The substrate mixture is first transferred to a well of a microtiter plate. The translation mixture is then carefully applied underneath the substrate mixture to form a bilayer. In this method, diffusion of both the translation mixture and substrate mixture facilitates the continuous supply of substrates and continuous removal of small by-products

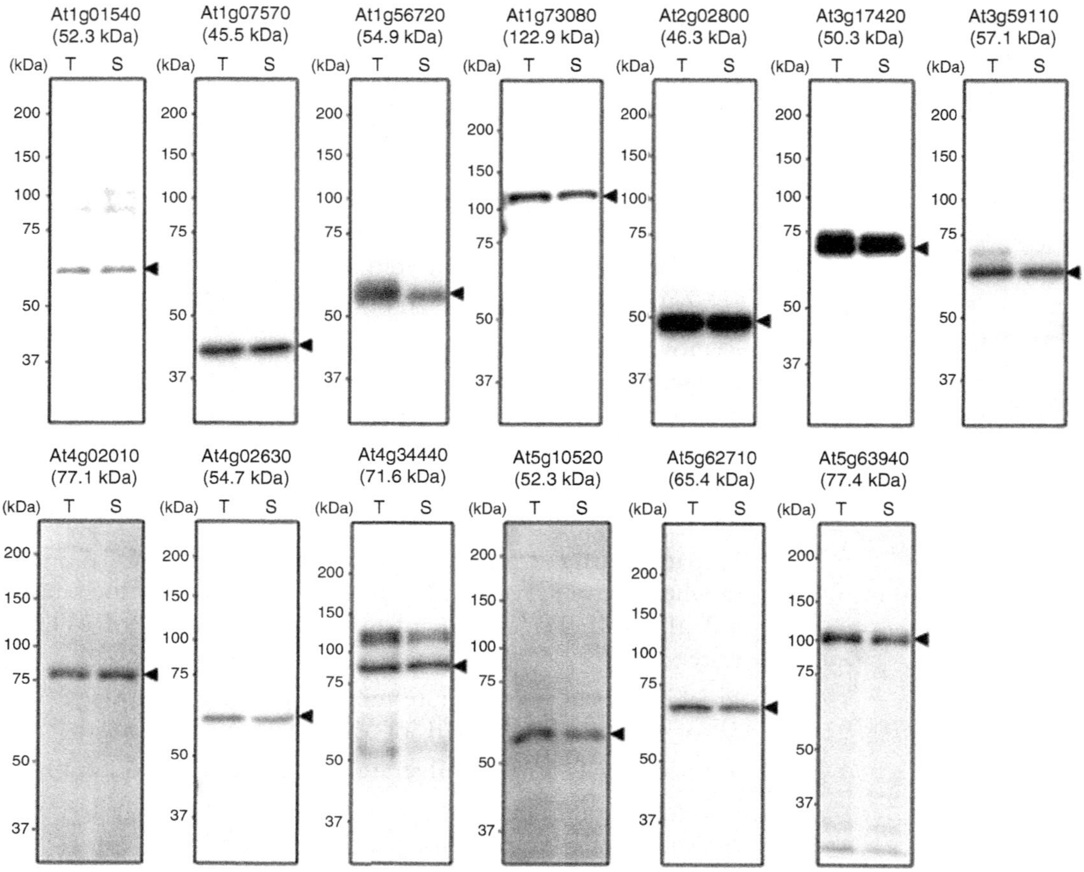

Fig. 4 Synthesis of *Arabidopsis* receptor kinases using the wheat germ cell-free system. Transcription templates for 16 types of *Arabidopsis* receptor kinases attached to the His-tag at the N terminus were prepared using the split-primer PCR method. Using these templates, the *Arabidopsis* receptor kinases were synthesized by the bilayer cell-free synthesis method. After the synthesis reaction, the total reaction mixtures were divided into supernatant fractions and pellet fractions by centrifugation. The total reaction mixtures (T) and the supernatant fractions (S) were subjected to SDS-PAGE, and the synthesized receptor kinases were detected by Western blot analysis using an anti-His-tag antibody. The predicted molecular sizes of the receptor kinases are shown in *parentheses*. *Arrowheads* indicate the synthesized receptor kinases. All receptor kinases are synthesized in the soluble fraction

7. Incubate the plate at 26 °C for 16 h.
8. To verify synthesis of the plant receptor kinases, analyze the reaction mixture by sodium dodecyl sulfate–polyacrylamide gel electrophoresis and Western blot analysis (Fig. 4).

4 Notes

1. We generally use pEU plasmids for wheat germ cell-free protein synthesis. pEU plasmids have SP6 promoter sequence and omega sequence or E01 sequence in upstream of multiple cloning site [6, 7]. SP6 promoter sequence is necessary for

in vitro transcription and omega sequence or E01 sequence stimulates initiation of translation of uncapped mRNAs. When using these plasmids, primer sets described in our previous reports [7, 16] can be used for the preparation of transcription templates by PCR.

2. By using a two-step PCR, the SP6 promoter sequence, translation enhancer sequence (E02), and 3′-UTR sequence are attached to the ORF sequence of a target gene. In the second PCR, we use two types of 5′ primer. By dividing the SP6 promoter sequence into two primers, we prevent amplification of nonspecific PCR fragments not suitable for protein production. In addition, a tag fragment can be attached to the N terminus of the target protein by exchange with the deSP6 primer sequence. In this case, the coding sequence of the tag fragment is added into the N~N region of the primer shown below. This primer is used instead of the deSP6 primer in split-primer PCR (5′-GGTGACACTATAGAACTCACCTATCTC TCTACACAAAACATTTCCCTACATAC AACTTTCAACTTCCTATTATGN~NCCACCCACCACC ACCAATG-3′).
3. If a weak band is obtained in the first PCR, 5 μL of the first PCR products should be added to the second PCR. In this case, the first PCR products need to be treated with exonuclease I in order to remove the primers contained in the first PCR mixture. Add 0.5 μL of exonuclease I (5 units/μL) to 50 μL of the first PCR mixture, and incubate at 37 °C. After 30 min, inactivate exonuclease I by heat treatment (80 °C, 30 min.), and use 5 μL for the second PCR.

References

1. Marcus A, Feely J (1964) Activation of protein synthesis in imbibition phase of seed germination. Proc Natl Acad Sci U S A 51:1075–1079
2. Spirin AS, Baranov VI, Ryabova LA, Ovodov SY, Alakhov YB (1988) A continuous cell-free translation system capable of producing polypeptide in high yield. Science 242:1162–1164
3. Kawarasaki Y, Kawai T, Nakano H, Yamane T (1995) A long-lived batch reaction system of cell-free protein synthesis. Anal Biochem 226:320–324
4. Nakano H, Tanaka T, Kawarasaki Y, Yamane T (1996) Highly productive cell-free protein synthesis system using condensed wheat-germ extract. J Biotechnol 46:275–282
5. Madin K, Sawasaki T, Ogasawara T, Endo Y (2000) A highly efficient and robust cell-free protein synthesis system prepared from wheat embryos: Plants apparently contain a suicide system directed at ribosomes. Proc Natl Acad Sci U S A 97:559–564
6. Kamura N, Sawasaki T, Kasahara Y, Takai K, Endo Y (2005) Selection of 5′-untranslated sequences that enhance initiation of translation of a cell-free protein synthesis system from wheat embryos. Bioorg Med Chem Lett 15:5402–5406
7. Sawasaki T, Ogasawara T, Morishita R, Endo Y (2002) A cell-free protein synthesis system for high-throughput proteomics. Proc Natl Acad Sci U S A 99:14652–14657
8. Sawasaki T, Hasegawa Y, Tsuchimochi M, Kamura N, Ogasawara T, Kuroita T, Endo Y (2002) A bilayer cell-free protein synthesis sys-

tem for high-throughput screening of gene products. FEBS Lett 514:102–105

9. Vinarov DA, Lytle BL, Peterson FC, Tyler EM, Volkman BF, Markley JL (2004) Cell-free protein production and labeling protocol for NMR-based structural proteomics. Nat Methods 1:149–153
10. Nemoto K, Takemori N, Seki M, Shinozaki K, Sawasaki T (2015) Members of the plant CRK superfamily are capable of trans- and autophosphorylation of tyrosine residues. J Biol Chem 290:16665–16677
11. Gish LA, Clark SE (2011) The RLK/Pelle family of kinases. Plant J 66:117–127
12. Morillo SA, Tax FE (2006) Functional analysis of receptor-like kinases in monocots and dicots. Curr Opin Plant Biol 9:460–469
13. The Arabidopsis Genome Initiative (2000) Analysis of the genome sequence of the flowering plant *Arabidopsis thaliana*. Nature 408:796–815
14. Dardick C, Chen J, Richter T, Ouyang S, Ronald P (2007) The rice kinase database. A phylogenetic database for the rice kinome. Plant Physiol 143:579–586
15. Shiu SH, Bleecker AB (2003) Expansion of the receptor-like kinase/Pelle gene family and receptor-like proteins in Arabidopsis. Plant Physiol 132:530–543
16. Nemoto K, Seto T, Takahashi H, Nozawa A, Seki M, Shinozaki K, Endo Y, Sawasaki T (2011) Autophosphorylation profiling of Arabidopsis protein kinases using the cell-free system. Phytochemistry 72:1136–1144

Chapter 5

Purification of Plant Receptor Kinases from Plant Plasma Membranes

Jin Suk Lee

Abstract

Receptor kinases play a central role in various biological processes, but due to their low abundance and highly hydrophobic and dynamic nature, only a few of them have been functionally characterized, and their partners and ligands remain unidentified. Receptor protein extraction and purification from plant tissues is one of the most challenging steps for the success of various biochemical analyses to characterize their function. Immunoprecipitation is a widely used and selective method for enriching or purifying a specific protein. Here we describe two different optimized protein purification protocols, batch and on-chip immunoprecipitation, which efficiently isolate plant membrane receptor kinases for functional analysis.

Key words Receptor kinase, Purification, Protein-protein interaction, Immunoprecipitation, Biosensor chip

1 Introduction

The plasma membrane is a complex structure that separates the cell from its external environment while providing a means for cellular communication. Membrane-bound receptor kinases are composed of an ectodomain generally involved in ligand binding, a single pass transmembrane domain, and an intracellular kinase domain. There are more than six hundred receptor kinases encoded by the *Arabidopsis* genome which coordinate extracellular signals to regulate numerous biological processes including growth, development, and defense [1–7]. However, despite their importance in biology, it is not fully understood how receptor kinases form a signaling complex at the plasma membrane. Furthermore, among a handful of functionally characterized receptor kinases, their partners and ligands remain unidentified. Their low natural abundance, instability in nonnative environments, and highly hydrophobic and dynamic nature are factors that have complicated the research process and contributed to the lack of understanding. The functional study of receptor kinases starts with protein extraction and

Reidunn Birgitta Aalen (ed.), *Plant Receptor Kinases: Methods and Protocols*, Methods in Molecular Biology, vol. 1621,
DOI 10.1007/978-1-4939-7063-6_5,

purification from a complex mixture (such as cell lysates) followed by various biochemical and proteomics studies. To be able to conduct functional analysis, it is therefore important to use a proper method to enrich or purify a protein of interest. Immunoprecipitation (IP) is a powerful separation technique, widely used to purify a protein [8]. However, because of the complexities associated with studying a membrane receptor kinase *in planta*, there are several variables and factors that should be considered when designing an IP experiment to isolate adequate amounts and successfully purify this protein.

The ERECTA-family leucine-rich repeat (LRR) receptor kinases, ERECTA, ERL1, and ERL2, control a wide range of biological processes, including organ growth, reproductive tissue development, immunity, and stomatal patterning [4, 9, 10]. Several *EPIDERMAL PATTERNING FACTOR* (*EPF*) family members that encode small cysteine-rich peptides have recently been discovered as signals for the ERECTA-family to control two different biological processes: stomata development and aboveground growth [11–15]. The methods described below were used in studying ligand-receptor and receptor-receptor interactions, as well as their binding affinity. Co-IP assays using solubilized microsomal fractions, which were isolated from double-transgenic *Arabidopsis* plants expressing pairwise combinations of epitope-tagged receptors, demonstrated that the ERECTA-family forms receptor homodimers as well as heterodimers with the LRR protein TOO MANY MOUTHS (TMM), while TMM does not homodimerize in vivo [12]. In addition to these receptor-receptor interactions, Co-IP or IP of ERECTA-GFP from purified membrane fractions, followed by biochemical studies, was used to detect associations of EPF ligands and ERECTA receptor kinase [11–13]. Furthermore, identification of EPF ligands as well as optimized receptor purification methods successfully provided the experimental setups to discover that two contrasting peptides, EPF2 and STOMAGEN/EPFL9, compete for binding to the same receptor kinase, ERECTA, to fine-tune stomatal patterning [13]. Here we describe two different protein purification protocols, batch and on-chip IP, from plants either stably or transiently expressing tagged proteins, to prepare membrane receptor kinases efficiently for various biochemical analyses in order to study their function. These methods should also be readily applicable to other membrane receptor kinases in plants.

2 Materials

2.1 Purification of Microsomal Membrane Fractions

1. Mortars and pestles.
2. Sonicator.
3. Empty chromatography columns (731-1550, Bio-Rad).

4. Ultracentrifuge capable of 100,000 × *g* at 4 °C.
5. Homogenization buffer: 100 mM Tris–HCl pH 8.8, 150 mM NaCl, 1 mM EDTA, 20% glycerol, 0.75% polyvinylpolypyrrolidone (PVPP), 20 mM NaF, 2 mM Na_3VO_4, 1 mM PMSF, and 1 complete protease inhibitor cocktail (Roche) per 50 mL extraction buffer.

   Make buffers fresh using premade stock solutions.

   Add solid insoluble PVPP to the extraction buffer 1–2 h before use; allow hydrating fully (*see* **Note 1**).
6. 100 mM phenylmethanesulfonyl fluoride (PMSF) in isopropanol.
7. YEB medium: beef extract (5 g/L), yeast extract (1 g/L), peptone (5 g/L), sucrose (5 g/L), and $MgSO_4 \cdot 7H_2O$ (0.5 g/L), pH 7.0.
8. Infiltration medium: 10 mM $MgCl_2$, 10 mM MES (pH 5.6), and 150 μM acetosyringone.

   Make buffers fresh using premade stock solutions.

2.2 Membrane Protein Extraction

1. Sonicator.
2. Membrane solubilization buffer: 100 mM Tris–HCl at pH 7.3, 150 mM NaCl, 1 mM EDTA, 10% glycerol, 1% Triton X-100, 20 mM NaF, 1 mM PMSF, and 1 complete protease inhibitor cocktail (Roche) per 50 mL extraction buffer.
3. Ultracentrifuge capable of 100,000 × *g* at 4 °C.
4. Bradford Protein Assay Kit (Bio-Rad).

2.3 Immunoprecipitation

1. Magnet separator (Invitrogen).
2. Mixer allowing tilting and rotation of tubes.
3. Dynabeads Protein G (Invitrogen).
4. Anti-GFP (ab290, Abcam).
5. Phosphate buffer saline (PBS) pH 7.4 with and without 0.1% Tween 20.

2.4 On-chip Receptor Protein Purification

1. Biosensor chip.
2. UV plasma.
3. Plate shaker.
4. Anti-GFP (ab290, Abcam).
5. PBS, pH 7.4.
6. PBS with 2% bovine serum albumin (BSA).

3 Methods

3.1 Preparing Plant Material

Arabidopsis Seedlings

1. *Arabidopsis* seedlings, expressing full-length receptor with an epitope tag, are obtained by sterilizing seeds in 33% bleach solution with 0.1% Triton for 10–12 min, followed by extensive washing with sterile water (*see* **Note 2**).
2. Sow seeds on ½ MS 1% sucrose agar plates and keep in a cold room in the dark for 2–3 days.
3. Grow them for 12 days in a regular growth room (16 h light/8 h dark cycle, temperature of 22 °C) and freeze samples in liquid nitrogen (*see* **Note 3**).
4. Grind frozen tissue to fine powder with prechilled mortar and pestle in liquid nitrogen.
5. Immediately transfer the finely ground powder into two 1.5 mL microcentrifuge tubes precooled in liquid nitrogen. The volume of the powder in each 1.5 mL tube must be around 0.5 mL.
6. The powder sample can be immediately used or stored in a −80 °C freezer until protein extraction (*see* **Note 4**).

Nicotiana benthamiana Leaves

1. Inoculate single colony of *Agrobacterium* that has been previously transformed with the epitope-tagged receptor and silencing suppressor p19 expression clone in 3 mL LB with appropriate antibiotics, respectively. Grow overnight at 28 °C, 200 rpm.
2. Use 1 mL of the overnight culture to inoculate 20 mL YEB medium with same antibiotics, plus 150 μM acetosyringone and 10 mM MES (pH 5.6) (*see* **Note 5**) and grow overnight at 28 °C, 200 rpm.
3. Precipitate the bacteria culture (5000 × *g*, 15 min at room temperature) and resuspend the pellet in 3 mL of the infiltration medium (*see* **Note 6**).
4. Measure the absorbance at 600 nm of a 1 in 10 dilution of resuspended culture and adjust the final OD_{600} to 1 with infiltration medium.
5. Leave on the bench at room temperature for 4–5 h before infiltration.
6. Just before infiltration, mix *Agrobacterium* suspensions that allow expression of p19 and receptor to have an optical density at 600 nm of 0.3 for each strain.
7. The mixed bacterial suspensions were then infiltrated into young, but fully expanded, leaves of tobacco plants using a needleless 1 mL syringe (*see* **Note 7**).

8. After infiltration, incubate plants under normal growing conditions (16 h light/8 h dark cycle, temperature of 22 °C) for 2 days to generate transgenic tobacco plants transiently expressing tagged receptor kinase (*see* **Note 8**) and freeze leaf tissues within the infiltrated zones in liquid nitrogen.
9. Grind frozen tissue to fine powder with prechilled mortar and pestle in liquid nitrogen.
10. Immediately transfer the powder into 50 mL tubes precooled in liquid nitrogen. The volume of the powder in each 50 mL tube must be around 5 mL.
11. Store the powder sample in a −80 °C freezer until protein extraction.

3.2 Purification of Microsomal Membrane Fraction

Arabidopsis Microsomal Membrane Fraction

1. Prepare fresh extraction buffer with presoaked 0.75% PVPP to remove phenolic compounds (1 mL per 0.5 mL volume of the tissue powder).
2. Add two volumes of cold homogenization buffer to the tube with ground powder and vortex immediately so that all the ground powder is quickly mixed with extraction buffer. The tissue should never be allowed to thaw.
3. Incubate at 4 °C for 10 min. with gentle mixing.
4. Centrifuge lysate at 10,000 × *g* at 4 °C for 10 min. This step may be repeated if necessary.
5. Transfer the supernatant to a new tube to remove cell debris and PVPP.
6. Perform ultracentrifugation at 100,000 × *g* for 30 min at 4 °C.
7. Remove the supernatant containing soluble proteins. The compact pellet obtained contains all of the cell's microsomal fraction.

Tobacco Microsomal Membrane Fractions

1. Prepare fresh extraction buffer with presoaked 0.75% PVPP to remove phenolic compounds (15 mL per 5 mL volume of the tissue powder).
2. Add three volumes of cold homogenization buffer to the tube with ground powder and vortex immediately so that all the ground powder is quickly mixed with extraction buffer. The tissue should never be allowed to thaw to prevent degradation.
3. Incubate at 4 °C for 10 min. with gentle mixing.
4. Centrifuge lysate at 10,000 × *g* at 4 °C for 10 min.
5. Transfer the supernatant to a new tube and then pass the supernatant through an empty column (Bio-Rad) to trap cell debris and PVPP (*see* **Note 9**).
6. Sonicate the flow through using two 10-s pulses (30 s in between pulses) using a probe sonicator (*see* **Note 10**).

7. Perform ultracentrifugaton at 100,000 × *g* for 30 min at 4 °C.
8. Remove the supernatant containing soluble proteins. The compact pellet obtained contains all of the cell's microsomal fraction.

3.3 Membrane Protein Extraction

1. Add 1 mL membrane solubilization buffer for 1 g of *Arabidopsis* samples and 2 ml membrane solubilization buffer for 5 g of tobacco samples and then resuspend well to bring the pellet into solution.
2. Sonicate the sample using two 10s pulses (30 s in between pulses) using a probe sonicator and allow the extract to incubate with the buffer for at least 30 min with gentle mixing at 4 °C.
3. Perform ultracentrifugation at 100,000 × *g* for 30 min at 4 °C to remove any insoluble particles.
4. Transfer the supernatant to a fresh tube and measure the protein amount using the Bradfrod assay kit.
5. Adjust and transfer 2–2.5 mg/mL protein to a new tube and keep on ice in the cold room for future use. Take a small aliquot to use as input and mix with 4× sample buffer.

3.4 Immunoprecipitation

1. Resuspend Dynabeads Protein G (Invitrogen) in the vial with rotation for 5 min.
2. Transfer resuspended 25 μL Dynabeads Protein G to a 1.5 mL microcentrifuge tube.
3. Place the tube on the magnet stand to separate the beads from the solution, and remove the supernatant.
4. Resuspend with 200 μL PBS by gentle pipetting.
5. Place the tube on the magnet stand to separate the beads from the solution, and remove the supernatant.
6. Remove the tube from the magnet stand and add 1 μL anti-GFP (Abcam) diluted in 200 μL PBS with Tween 20.
7. Incubate with rotation overnight at 4 °C.
8. The next day, place the tube on the magnetic stand and remove the supernatant.
9. Resuspend with 200 μL PBS with Tween 20 by gentle pipetting twice.
10. Add 2–2.5 mg/mL protein extract from Subheading 3.3, **step 5** to the tube and incubate with rotation for 1–2 h at 4 °C (*see* **Note 11**).
11. Place the tube on the magnet stand to separate the beads from the solution, and remove the supernatant.
12. Wash at least 3–4 times using 500 μL PBS without or with Tween 20 for each wash.

13. Resuspend with 200 μL PBS and transfer the bead suspension to a clean 1.5 mL tube. This is recommended to avoid co-elution of proteins bound to the tube wall.
14. Place the tube on the magnet stand to separate the beads from the solution, and remove the supernatant.
15. At this point, the receptor kinase is purified by IP and can be used for interaction studies with other proteins such as peptide ligands (*see* **Note 12**).

3.5 On-chip Receptor Protein Purification

1. Place the biosensor chip into 1:1 methanol/acetone solution and sonicate the chip in methanol:acetone mixture for 5 min in sonication bath.
2. Take out the biosensor chip from methanol:acetone mixture and remove the liquid from the surface of the chip by blotting the chip in paper Kim wipes (*see* **Note 13**).
3. Place the biosensor chip into isopropyl alcohol solution and sonicate the chip in isopropyl alcohol for 5 min in sonication bath. Remove the liquid from the surface of the chip by blotting the chip in paper Kim wipes.
4. Place the biosensor chip into distilled water and sonicate the chip for 5 min in sonication bath. Remove the liquid from the surface of the chip by blotting the chip in paper Kim wipes.
5. Repeat **step 3** and sonicate the biosensor chip in distilled water for 5 min in sonication bath. Remove the liquid from the surface of the chip by blotting the chip in paper Kim wipes.
6. Gently dry the cleaned biosensor chip with nitrogen.
7. Treat the dry biosensor chip with UV plasma for 15 min.
8. Prepare 2 μL anti-GFP (Abcam) solution diluted in 100 μL PBS.
9. Place a drop of anti-GFP solution onto a clean biosensor chip and incubate at 4 °C in a humidity-controlled environment overnight (~16 h) (*see* **Note 14**).
10. Rinse the biosensor chip with 10 mL of sterile PBS by dilution rinse in standard plastic 24-well plate (*see* **Note 15**).
11. Place 1 mL of PBS with 2% BSA to the biosensor chip in the well in standard plastic 24-well plate. Block the biosensor chip surface for 2 h at 4 °C in a humidity-controlled environment shaking.
12. Perform the dilution rinse as described in **step 3**.
13. Place 1 mL of 2–2.5 mg/mL protein extract from Subheading 3.3, **step 5** to the biosensor chip in the well in standard plastic 24-well plate, and incubate for 4 h at 4 °C in a humidity-controlled environment shaking.

14. Rinse the receptor-modified biosensor chips with 10 mL of sterile PBS by dilution rinse in standard plastic 24-well plate as described in **step 3**.
15. Gently dry the rinsed receptor-modified biosensor chip with nitrogen to remove the liquid, and place the receptor-modified biosensor chip onto a biosensor platform (*see* **Note 16**). At this point, the biosensor chip is immobilized with purified receptor kinase and can be used for investigating binding affinity and kinetics of the ligand-receptor pair [13, 16].

4 Notes

1. Using PVPP is preferable instead of polyvinylpyrrolidone (PVP) for phenolic compound removal because it can be easily removed from the sample at the filtration and centrifugation steps.
2. We recommend generating stable *Arabidopsis* plants expressing epitope-tagged full-length receptor kinase by using genomic DNA constructs driven by their own promoters. It is important to determine if an epitope-tagged receptor is functional by checking its ability to complement the respective mutant phenotype and by checking that it is expressed in the correct cellular location. Note that we also use the kinase-deleted version of an epitope-tagged receptor kinase, especially for the interaction studies with peptide ligands since the kinase domain of receptors is not required for ligand binding and removal of the kinase domain increases receptor stability [12, 17–19].
3. Densely sow seeds on one 150 mm plate to prepare approximately 1 g (fresh weight) of 12-day-old *Arabidopsis* seedlings required for one IP.
4. We have stored the powder sample for 3–4 months without any apparent protein degradation. To perform IP of several samples at the same time, it is practical to grind samples in advance and keep them in a −80 °C freezer.
5. This is generally more than enough, but if you have many transformations to do, use more.
6. **Step 3** can be repeated to remove traces of antibiotic, which will kill the leaf tissue after infiltration.
7. The rates of transformation are dependent on the viability of the plant and *Agrobacterium*. Therefore, it is important to repeat infiltrations in two midrib regions on 2–3 leaves in the same plant and use at least 6–8 independent healthy plants.
8. The expression of sufficient amounts of receptor kinase is one of the most important steps for the success of various biochemical

analysis including interaction studies of receptor kinases and peptide ligands. As such, we also use tobacco (*Nicotiana benthamiana*) plants that transiently express epitope-tagged receptors instead of *Arabidopsis* lines to isolate microsomal fractions for high-efficiency expression [11–13]. Proper subcellular localization and expression levels need to be monitored at a different time after infiltration to prevent the formation of aggregates which may cause false interactions [20]. It is also necessary to use good negative controls for following biochemical analysis. Two days after infiltration is generally a good time to harvest tobacco samples.

9. If the flow rate is slow, you can pipet up and down several times as necessary. In addition to the centrifugation step, this extra step is required to remove cell wall debris and PVPP efficiently for tobacco leaves.
10. The sample should be on ice while performing sonication to prevent overheating; keep probes away from the sample-air interface to minimize foaming.
11. Each IP should contain the same protein concentration in order to obtain consistent results, and depending on the affinity of the antibody, it may be necessary to adjust incubation times for optimal binding.
12. For the following biochemical analysis involving an interaction study with peptide ligands, we add peptides in the binding buffer (50 mM MES-KOH, pH 5.5 with 100 mM sucrose) to the purified receptor by IP and incubate with rotation for 0.5–1 h at 4 °C. Optimization of incubation time, binding, and washing conditions need to be examined extensively to prevent false protein-protein interactions using good positive and negative controls.
13. Do not touch the chip surface with paper to remove the liquid. Tap gently the edge of the biosensor chip with paper Kim wipes.
14. In drop incubation, make sure that the anti-GFP solution covers surface of the biosensor chip and that the incubation is done in humidity-controlled environment to prevent the liquid evaporation.
15. While doing dilution rinsing, always keep some liquid in wells and diminish surface drying to reduce the drying-induced antibody denaturation. Do not touch the chip surface to remove liquid.
16. Carefully remove all liquid from the gold electrode on the down-facing side of biosensor chip. Receptor coverage and alignment can be examined by using atomic force microscopy.

References

1. Torii KU (2012) Mix-and-match: ligand-receptor pairs in stomatal development and beyond. Trends Plant Sci 17(12):711–719
2. Hazak O, Hardtke CS (2016) CLAVATA 1-type receptors in plant development. J Exp Bot 67(16):4827–4833
3. Wu Y, Zhou JM (2013) Receptor-like kinases in plant innate immunity. J Integr Plant Biol 55(12):1271–1286
4. Shpak ED (2013) Diverse roles of ERECTA family genes in plant development. J Integr Plant Biol 55(12):1238–1250
5. Niederhuth CE, Cho SK, Seitz K, Walker JC (2013) Letting go is never easy: abscission and receptor-like protein kinases. J Integr Plant Biol 55(12):1251–1263. doi:10.1111/jipb.12116
6. De Smet I, Voss U, Jurgens G, Beeckman T (2009) Receptor-like kinases shape the plant. Nat Cell Biol 11(10):1166–1173
7. Kim BH, Kim SY, Nam KH (2013) Assessing the diverse functions of BAK1 and its homologs in Arabidopsis, beyond BR signaling and PTI responses. Mol Cells 35(1):7–16
8. Kaboord B, Perr M (2008) Isolation of proteins and protein complexes by immunoprecipitation. Methods Mol Biol 424:349–364
9. van Zanten M, Snoek LB, Proveniers MC, Peeters AJ (2009) The many functions of ERECTA. Trends Plant Sci 14(4):214–218
10. Bemis SM, Lee JS, Shpak ED, Torii KU (2013) Regulation of floral patterning and organ identity by Arabidopsis ERECTA-family receptor kinase genes. J Exp Bot 64(17):5323–5333
11. Uchida N, Lee JS, Horst RJ, Lai HH, Kajita R, Kakimoto T, Tasaka M, Torii KU (2012) Regulation of inflorescence architecture by intertissue layer ligand-receptor communication between endodermis and phloem. Proc Natl Acad Sci U S A 109(16):6337–6342
12. Lee JS, Kuroha T, Hnilova M, Khatayevich D, Kanaoka MM, McAbee JM, Sarikaya M, Tamerler C, Torii KU (2012) Direct interaction of ligand-receptor pairs specifying stomatal patterning. Genes Dev 26(2):126–136
13. Lee JS, Hnilova M, Maes M, Lin YC, Putarjunan A, Han SK, Avila J, Torii KU (2015) Competitive binding of antagonistic peptides fine-tunes stomatal patterning. Nature 522(7557):439–443
14. Abrash EB, Bergmann DC (2010) Regional specification of stomatal production by the putative ligand CHALLAH. Development 137(3):447–455
15. Abrash EB, Davies KA, Bergmann DC (2011) Generation of signaling specificity in Arabidopsis by spatially restricted buffering of ligand-receptor interactions. Plant Cell 23(8):2864–2879
16. Takayama S, Shimosato H, Shiba H, Funato M, Che FS, Watanabe M, Iwano M, Isogai A (2001) Direct ligand-receptor complex interaction controls *Brassica* self-incompatibility. Nature 413(6855):534–538
17. Shpak ED, Lakeman MB, Torii KU (2003) Dominant-negative receptor uncovers redundancy in the Arabidopsis ERECTA Leucine-rich repeat receptor-like kinase signaling pathway that regulates organ shape. Plant Cell 15(5):1095–1110
18. Ogawa M, Shinohara H, Sakagami Y, Matsubayashi Y (2008) Arabidopsis CLV3 peptide directly binds CLV1 ectodomain. Science 319(5861):294
19. Shinohara H, Ogawa M, Sakagami Y, Matsubayashi Y (2007) Identification of ligand binding site of phytosulfokine receptor by on-column photoaffinity labeling. J Biol Chem 282(1):124–131
20. Bleckmann A, Weidtkamp-Peters S, Seidel CA, Simon R (2010) Stem cell signaling in Arabidopsis requires CRN to localize CLV2 to the plasma membrane. Plant Physiol 152(1):166–176

Part II

Ligand Binding and Activation of PRKs

Chapter 6

Photoaffinity Labeling of Plant Receptor Kinases

Hidefumi Shinohara and Yoshikatsu Matsubayashi

Abstract

Defining the ligand-binding activity of receptors is important because the binding of ligands is the initial reaction in secreted ligand-dependent cell-to-cell communication. Photoaffinity labeling is one of the most efficient biochemical techniques for detecting direct ligand-receptor interactions. Here, we describe photoaffinity labeling to visualize the direct interaction between peptide ligands and their receptors by using photoactivatable and radioactive peptide ligand derivatives.

Key words Photoaffinity labeling, Receptor kinase, Peptide ligand, Ligand-receptor interaction

1 Introduction

In general, receptor kinases are functionally divided into two groups. One group is involved in direct ligand binding, and defined members of this group are called "direct receptors." Members of the other group are termed "co-receptors" and are not involved in direct ligand binding but are required for ligand-dependent signal transduction. It is important to define the ligand-binding activity of receptors because the binding of ligands to their receptors is the initial reaction that occurs in secreted ligand-dependent cell-to-cell communication.

Photoaffinity labeling is one of the most efficient biochemical techniques for detecting direct ligand-receptor interactions. This method is commonly used for the detection of direct interactions between peptide ligands and their specific receptors, e.g., CLV1-CLV3 [1], BAM1/2-CLV3/CLE9 [2], TDR-TDIF [3], SKM1-CLE45 [4], CEPR-CEP [5], RGFR-RGF [6], PEPR-AtPep [7] and GSO1/SGN3/GSO2-CIF [8]. We first prepared photoreactive ligand molecules containing radiolabeled photoactivatable 4-azidosalicylic acid. With ultraviolet (UV) light irradiation, the azide group of the photoreactive ligand photolyzes to form a reactive nitrene, which forms a covalent bond with proximal amino acid residues within the receptor proteins. A cross-linked

Reidunn Birgitta Aalen (ed.), *Plant Receptor Kinases: Methods and Protocols*, Methods in Molecular Biology, vol. 1621,
DOI 10.1007/978-1-4939-7063-6_6,

ligand-receptor complex can be detected by SDS-PAGE and autoradiography. Such labeling of the receptors provides strong and direct evidence of ligand-receptor interactions.

Here, we present photoaffinity labeling to visualize the direct interaction between peptide ligands and their receptors. We introduce information on (1) how to prepare the photoreactive peptide derivative, (2) how to introduce radioactive iodine into the peptide, and (3) how to detect an interaction between the peptide ligand and its receptor using photoaffinity labeling.

2 Materials

2.1 Preparation of Photoreactive 4-Azidosalicylic Acid (ASA)-Marked Peptide

1. [Lys]-substituted and Fmoc-protected peptide derivative (e.g., Fmoc-[Lys]^5-RGF1) (*see* **Notes 1–3**).
2. Acetonitrile.
3. $NaHCO_3$.
4. *N*-hydroxysuccinimidyl-4-azido salicylic acid (NHS-ASA, PIERCE, 27714).
5. Spin evaporator with vacuum pump.
6. Deprotection solution: 50% piperidine, 25% CH_3CN, and 25% water (v/v).
7. Freeze dryer.
8. Centrifuge.
9. HPLC system (JASCO International Co., model: PU-2080 pump combined with a DG-2080-53 degasser, an MX-2080-32 dynamic mixer, and a UV-2075 detector).
10. Chromatogram integrator (LAsoft, CDS plus).
11. Cadenza CD-C18 (3 μm) column (4.6 × 150 mm) (Imtakt CD005).
12. HPLC grade water containing 0.1% CH_3COONH_4.
13. HPLC grade 50% acetonitrile containing 0.1% CH_3COONH_4.

2.2 Radioactive Iodination of ASA-Marked Peptide

1. 0.8 M phosphate buffer and pH 7.5.
2. 0.4 mM ASA-marked peptide solution.
3. 0.47 mM NaI solution.
4. Isolation glove box.
5. Na[^{125}I] (PerkinElmer, NEZ033A, 37 MBq eq.) (*see* **Note 4**).
6. 0.4 mM chloramine-T solution (*see* **Note 5**).
7. 10% trifluoroacetic acid (TFA) (v/v).
8. HPLC system (JASCO International Co., model: PU-2080 pump combined with a DG-2080-53 degasser, an MX-2080-32 dynamic mixer, and a UV-2075 detector (λ; 220 nm)).

9. Cadenza CD-C18 (3 μm) column (4.6 × 150 mm) (Imtakt CD005).
10. Chromatogram integrator (LAsoft, CDS plus).
11. HPLC grade water containing 0.1% TFA.
12. HPLC grade acetonitrile containing 0.1% TFA.
13. Dose calibrator (AcroBio, CRC-25R).

2.3 Photoaffinity Labeling

1. Binding buffer: 50 mM MES-KOH, pH 5.5, and 100 mM sucrose (*see* **Note 6**).
2. [^{125}I]ASA-conjugated peptide solution.
3. 100 μM unlabeled peptide solution (as a competitor).
4. Wash buffer: 50 mM MES-KOH, pH 5.5, and 500 mM sucrose (*see* **Note 6**).
5. Ultracentrifuge.
6. UV lamp (model ENF-260C/J Spectronics Co. Ltd).
7. Solubilization buffer: 20 mM HEPES-KOH, pH 7.5, 150 mM KCl, and 1% Triton X-100 (v/v).
8. Rotator.
9. Chamber or refrigerator set at 4 °C.
10. Anti-HaloTag pAb (Promega, G9281).
11. nProteinA Sepharose (GE, 17528001).
12. 2× SDS sample buffer: 0.1 M Tris–HCl, pH 6.8, 4% SDS, 12% ß-mercaptoethanol, 8% glycerol, and 0.02% bromophenol blue (v/v).
13. Heat block set at 95 °C.
14. 7.5% acrylamide gel for SDS-PAGE.
15. SDS-PAGE running buffer: 25 mM Tris, pH 8.2, 0.192 M glycine, and 0.1% SDS.
16. Imaging plate (Fujifilm, MS 2025).
17. Imaging plate cassette.
18. Imaging analyzer (GE, FLA Typhoon 9000).

3 Methods

3.1 Preparation of Photoreactive 4-Azidosalicylic Acid (ASA)-Marked Peptide

In this section, we describe the introduction of the photoreactive residue in target peptides using an NHS ester-activated cross-linker. The [Lys]-substituted and Fmoc-protected peptide derivative RGF1 (Fmoc-[Lys]^5-RGF1) [6] is used as a model for this study (*see* Fig. 1). It takes approximately 2–3 days to complete the entire protocol described in this chapter.

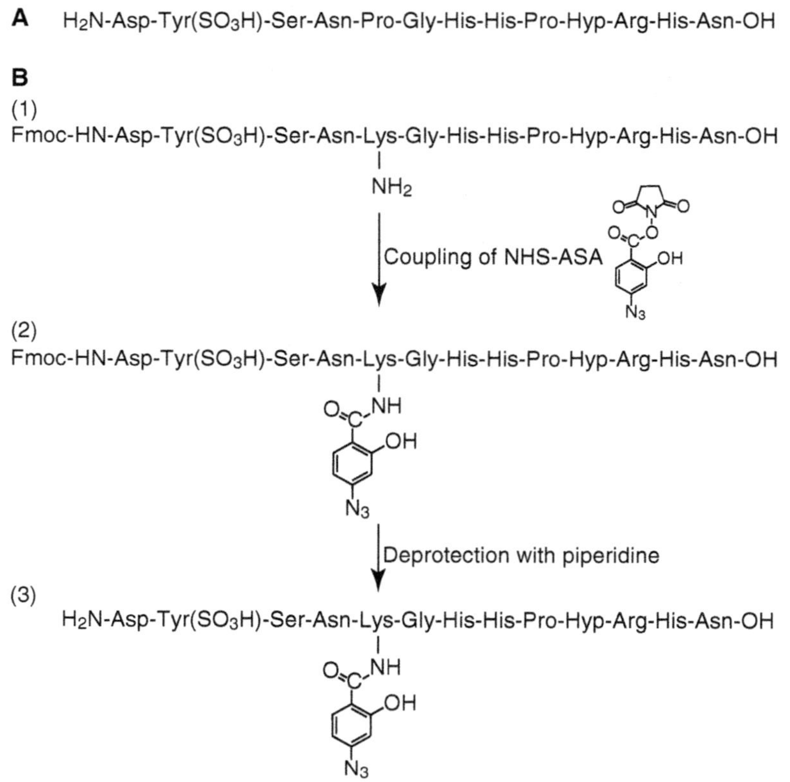

Fig. 1 (**a**) Structure of the RGF1 peptide. (**b**) Schematic outline of the preparation of ASA-marked RGF peptides; (1) Fmoc-[Lys^5]-RGF1, (2) Fmoc-ASA-RGF1, and (3) ASA-RGF1, respectively

1. Place 2.3 mg of Fmoc-[Lys^5]-RGF1 peptide powder (1.2 μmol eq.) into a 1.5 mL microtube.
2. Add 200 μL of 50% acetonitrile and dissolve the peptides.
3. Add 3 mg of $NaHCO_3$ and vortex. Check the pH of solution using a pH test paper. Confirm that the pH is higher than 8 (*see* **Note 7**).
4. Add 1.3 mg (4.8 μmol) of *N*-hydroxysuccinimidyl-4-azido salicylic acid (NHS-ASA) into the reaction solution. Vortex and sonicate to dissolve NHS-ASA powder.
5. Mix for 30 min in the dark at room temperature (*see* **Note 8**).
6. Set the reaction solution on a spin evaporator and evaporate the acetonitrile.
7. Add 200 μL of deprotection solution and vortex well.
8. Add 400 μL of water.
9. Lyophilize the sample (*see* **Note 9**).

10. Add 400 μL of water and vortex well to dissolve ASA-conjugated peptide (*see* **Note 10**).
11. Centrifuge at 21,500 × *g* for 10 min at room temperature. Remove the precipitant containing insoluble compounds (*see* **Note 11**).
12. Set up the HPLC. The mobile phase is composed of HPLC grade water containing 0.1% CH_3COONH_4 (eluent A) and HPLC grade 50% acetonitrile containing 0.1% CH_3COONH_4 (eluent B). The HPLC flow rate is 1 mL/min, and the elution gradient was 0–80% B over 15 min.
13. Load aliquots of the reaction mixture (100 μL) on a Cadenza CD-C18 (3 μm) column.
14. Collect the peaks containing ASA-conjugated peptide.
15. Repeat **steps 13–14** four times.
16. Put all collected samples together in a single microtube.
17. Set the sample on a spin evaporator and evaporate acetonitrile (*see* **Note 12**).
18. Lyophilize the sample (*see* **Note 13**).
19. Weigh lyophilized peptide powder and dissolve in water to be a final concentration of 0.4 mM.

3.2 Radioactive Iodination of ASA-Marked Peptide

In this section, we describe the iodination of ASA-marked peptide derivatives with radioactive iodine using the chloramine-T method. The photoactivatable RGF1 derivative ASA-RGF1 is used as a model for this study (*see* Fig. 2).

1. Mix 5 μL of 0.8 M phosphate buffer, pH 7.5, 15 μL of 0.4 mM ASA-marked peptide, and 5 μL of 0.47 mM NaI solution in 1.5 mL microtubes.
2. Put the reaction mixture in the isolation glove box (*see* **Note 14**).
3. Add 5 μL of $Na[^{125}I]$ (18.5 MBq eq.) and mix.
4. Add 10 μL of 0.4 mM chloramine-T solution.
5. Mix and incubate for 5 min at room temperature.
6. Add 10 μL of 10% TFA to stop the reaction.
7. Set up the HPLC. The mobile phase is composed of HPLC grade water containing 0.1% TFA (eluent A) and HPLC grade acetonitrile containing 0.1% TFA (eluent B). The HPLC flow rate is 1 mL/min, and the elution gradient is 25–50% B over 15 min.
8. Load the reaction mixture on a Cadenza CD-C18 (3 μm) column.
9. Collect the peaks containing iodinated peptides (*see* Fig. 3, **Note 15**).

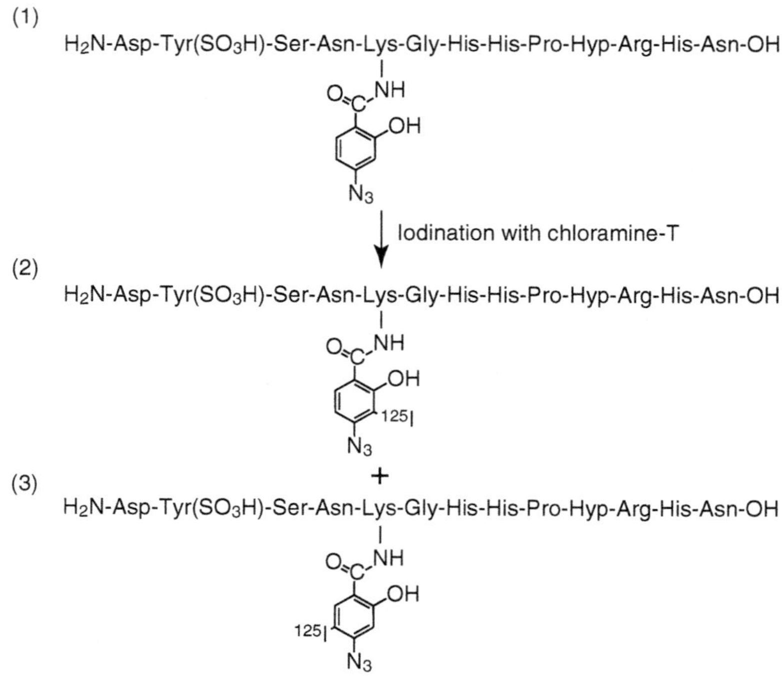

Fig. 2 Schematic outline of the radioactive iodination of ASA-marked RGF peptides; (1) ASA-RGF1, (2) monoiodinated [^{125}I]ASA-RGF1 (*ortho* position), and (3) monoiodinated [^{125}I]ASA-RGF1 (*para* position)

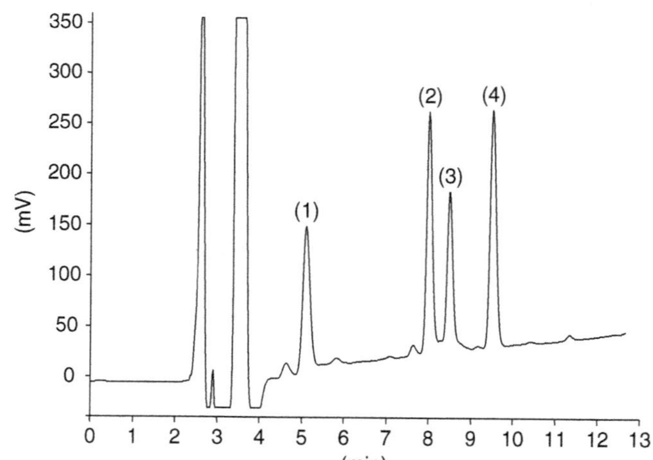

Fig. 3 Purification of iodinated peptides by HPLC on a reverse-phase column under the conditions described in step 7. Under this condition, two monoiodinated peptides ((2) and (3) in Fig. 2) were obtained at a retention time of 8.0 and 8.5 min (peaks 2 and 3), respectively. Peak 1, unlabeled ASA-RGF1; peak 4, chloramine-T

10. Check the volume of the collected sample.
11. Check the radiation dosage of the collected sample.
12. Calculate the amounts of iodinated peptides in the collected samples (*see* **Note 16**).
13. Calculate the concentrations and specific radioactivities of iodinated peptides (*see* **Note 17**).
14. Store at 4 °C (*see* **Note 18**).

3.3 Photoaffinity Labeling

In this section, we describe the method of photoaffinity labeling using radiolabeled and photoreactive peptide derivatives. The photoactivatable and radioactive RGF1 derivative [^{125}I]ASA-RGF1 is used as a model for this study.

1. Suspend 15 μL (≈750 μg) of the microsomal fraction derived from BY-2 cells expressing HaloTag-fused receptor kinases in 250 μL of the binding buffer (also refer to Chapter 3).
2. Add [^{125}I]ASA-conjugated peptide solution to a final concentration of 30 nM.
3. In a competition assay, add 2.5 μL of 100 μM unlabeled peptide solution to a final concentration of 10 μM as a competitor.
4. Mix and incubate for 30 min on ice.
5. Layer the reaction mixture onto 900 μL of wash buffer.
6. Centrifuge for 5 min at 100,000 × *g* and 4 °C (*see* **Note 19**).
7. Discard the supernatant.
8. Put the tubes with the remaining pellet on ice.
9. Direct an ultraviolet (UV) lamp on the samples and UV irradiate at 365 nm for 20 min.
10. Add 400 μL of solubilization buffer and sonicate to dissolve the pellet.
11. Rotate for 30 min at 4 °C.
12. Centrifuge for 30 min at 100,000 × *g* and 4 °C.
13. Put the supernatant into a new microtube and add 1 μL (1 μg eq.) of Anti-HaloTag pAb.
14. Rotate for 1 h at 4 °C.
15. Add 30 μL of nProteinA Sepharose beads.
16. Rotate for 1 h at 4 °C.
17. Centrifuge for 5 min at 3000 × *g* and 4 °C.
18. Discard the supernatant.
19. Add 400 μL of solubilization buffer and vortex.
20. Repeat **steps 16–19** three times.

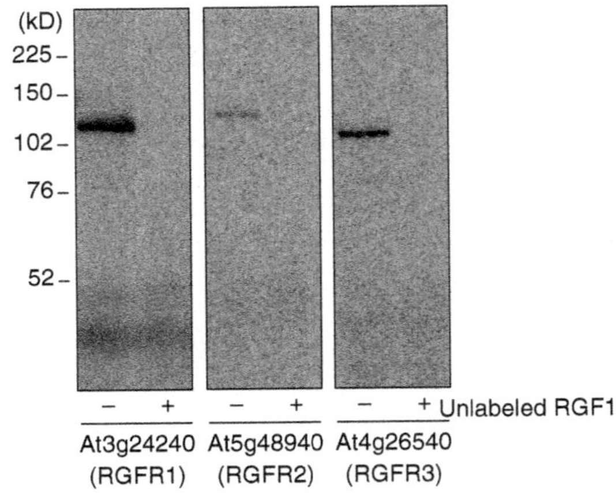

Fig. 4 Photoaffinity labeling using [^{125}I]ASA-RGF1 against membrane fractions derived from RGFR1/2/3 expressing BY-2 cells (*see* also Chapter 3). Competitive displacement of [^{125}I]ASA-RGF1 binding was detected by adding a 300-fold excess of unlabeled RGF1

21. Remove as much of the supernatant as possible.
22. Add 30 μL of 2× SDS sample buffer.
23. Heat at 95 °C for 5 min.
24. Centrifuge for 5 min at 3000 × *g* and 4 °C.
25. Load the supernatant of the heated sample on a 7.5% acrylamide gel and separate by SDS-PAGE.
26. After the SDS-PAGE, dry the gel with a gel dryer.
27. Place the dried gel onto an imaging plate (IP) cassette and put the IP and the dried gel in a dark room.
28. Expose the IP for the desired time.
29. Analyze the IP using a Typhoon FLA 9000 with a storage phosphor-scanning method (*see* Fig. 4).

4 Notes

1. The common techniques for labeling peptides involve the use of chemical groups that react with primary amines (-NH_2). Primary amines exist at the N-terminus of each peptide chain and in the side chain of lysine (Lys) amino acid residues. Target peptide derivatives with directly introduced photoreactive residues on the N-terminus amine or with additional Lys residues on the N- or C-terminal end of the peptides can be chosen. In some instances, N- or C-terminal modifications can interfere with the biological functions of the peptide. In these cases, [Lys]-substituted peptide derivatives should be prepared as described below.

2. Alanine-scanning experiments are required to determine the replaceable positions of amino acid residues in advance to prepare the [Lys]-substituted peptide derivatives. Alanine scanning is able to identify the specific amino acid residues responsible for a peptide's activity. For example, alanine substitution for Ser^3, Asn^4, Pro^5, Gly^6, His^7, His^8, Pro^9, and Arg^{11} within the 15-amino-acid sequence does not affect RGF bioactivity [6]. Lys should be replaced with residues that do not contribute to its bioactivity.
3. Fmoc residue for the protection of the N-terminal alpha amino group is required to prevent the coupling reaction of NHS-ASA with the N-terminal alpha amino group.
4. The half-life of ^{125}I is ≈60 days.
5. Fresh chloramine-T solution should be prepared for each experiment.
6. Once prepared, the suspension buffer can be stored at −20 °C.
7. An alkaline condition is important for the reaction with NHS-activated cross-linkers and for labeling compounds with primary amines.
8. ASA is a light-sensitive residue. Perform the following steps in a dark room as much as possible until cross-linker photoactivation has finished.
9. Lyophilization is complete after approximately 16 h.
10. The freeze-dried sample should be dissolved at once.
11. Released Fmoc residues are insoluble in water.
12. Acetonitrile evaporation is complete within 1 h.
13. Repeat this step until a dry peptide powder is obtained.
14. Perform the following experiments in a glove box. Handle ^{125}I radioactive compounds according to approved guidelines of your laboratory.
15. Under this condition, two monoiodinated (*ortho* or *para* position of the hydroxyl group, *see* Fig. 2) peptides are obtained (peak (2) and (3) in Fig. 3, respectively). There are some cases in which these two peptides have different binding affinities to their specific receptors, which likely depend on the distinct sequences of individual peptides. In the photoaffinity-labeling experiment, we recommend using both peptides as photoreactive ligands.
16. The amount of iodinated peptides is calculated by comparing the peak area of the collected sample with that of a known concentration of non-iodinated ASA-conjugated peptides.
17. Under this condition, the specific radioactivity of radioiodinated peptides is 50–200 Ci/mmol.

18. Do not freeze the sample because freezing results in the subsequent volatilization of radioiodine.
19. The receptor-bound and free [^{125}I]ASA-conjugated peptides are separated in this operation.

Acknowledgments

This research was supported by a Grant-in-Aid for Scientific Research (S) (25221105 to Y.M.), a Grant-in-Aid for Scientific Research on Innovative Areas (15H05957 to Y.M., 26113520 and 16H01234 to H.S.), and a Grant-in-Aid for Young Scientists (B) (25840111 to H.S.).

References

1. Ogawa M, Shinohara H, Sakagami Y, Matsubayashi Y (2008) Arabidopsis CLV3 peptide directly binds CLV1 ectodomain. Science 319:294
2. Shinohara H, Matsubayashi Y (2015) Reevaluation of the CLV3-receptor interaction in the shoot apical meristem: dissection of the CLV3 signaling pathway from a direct ligand-binding point of view. Plant J 82:328–336
3. Hirakawa Y, Shinohara H, Kondo Y, Inoue A, Nakanomyo I, Ogawa M, Sawa S, Ohashi-Ito K, Matsubayashi Y, Fukuda H (2008) Non-cell-autonomous control of vascular stem cell fate by a CLE peptide/receptor system. Proc Natl Acad Sci U S A 105:15208–15213
4. Endo S, Shinohara H, Matsubayashi Y, Fukuda H (2013) A novel pollen-pistil interaction conferring high-temperature tolerance during reproduction via CLE45 signaling. Curr Biol 23:1670–1676
5. Tabata R, Sumida K, Yoshii T, Ohyama K, Shinohara H, Matsubayashi Y (2014) Perception of root-derived peptides by shoot LRR-RKs mediates systemic N-demand signaling. Science 346:343–346
6. Shinohara H, Mori A, Yasue N, Sumida K, Matsubayashi Y (2016) Identification of three LRR-RKs involved in perception of root meristem growth factor in Arabidopsis. Proc Natl Acad Sci U S A 113:3897–3902.
7. Yamaguchi Y, Pearce G, Ryan C (2006) The cell surface leucine-rich repeat receptor for AtPep1, an endogenous peptide elicitor in Arabidopsis, is functional in transgenic tobacco cells. Proc Natl Acad Sci U S A 103:10104–101098.
8. Nakayama T, Shinohara H, Tanaka M, Baba K, Ogawa-Ohnishi M, Matsubayashi Y (2017) A peptide hormone required for Casparian strip diffusion-barrier formation in Arabidopsis roots. Science 355:284–286

Chapter 7

Quantitative Detection of Oxidative Burst upon Activation of Plant Receptor Kinases

Markus Albert and Ursula Fürst

Abstract

The oxidative burst or the production of reactive oxygen species (ROS) is a typical cellular response of both plants and animals to diverse abiotic and biotic stresses. Mainly, the (re-)active oxygen species include the superoxide anion (O_2^-), hydrogen peroxide (H_2O_2), and the hydroxyl radical (OH•). Here, we outline the detection of extracellularly produced ROS in plant leaf pieces using a chemiluminescence-based bioassay with the luminol L-012 as a substrate being oxidized in the presence of ROS. Since this type of assay is in use in many laboratories, e.g., as a readout for activation of plant receptor kinases, we include a discussion on the interpretation of results and points addressing problems with the buffers at suboptimal pH values that negatively influence the chemiluminescence production.

Key words Reactive oxygen species (ROS), Oxidative burst, Superoxide anion, Hydrogen peroxide, Chemiluminescence, Luminol, Biotic stress, Stress response, Plant development

1 Introduction

In living organisms, reactive oxygen species (ROS) are often produced intracellularly as by-products of electron transfer chains or due to oxidizing activity of diverse organelles such as peroxisomes, mitochondria, or chloroplasts [1]. Besides this, higher organisms also produce extra ROS that serve as important signaling molecules in diverse biological processes. In higher plants, extracellular ROS production seems essential during growth and developmental processes [2] or when plants are exposed to abiotic or biotic stresses [3–5]. The stress-related and autonomous ox-burst has been observed in plants first in 1983, when potato tubers were infected with *Phytophthora infestans*, a necrotrophic oomycete [6]. During plant–pathogen interaction, ROS production belongs to the "early" and fast responses when plant pathogens or microbe-associated molecular patterns are sensed by plant cell membrane-bound pattern recognition receptors [4]. This oxidative burst, dependent on ligand binding to the receptor and an active

Reidunn Birgitta Aalen (ed.), *Plant Receptor Kinases: Methods and Protocols*, Methods in Molecular Biology, vol. 1621, DOI 10.1007/978-1-4939-7063-6_7, © Springer Science+Business Media LLC 2017

intracellular kinase domain, is measureable about 2–20 min after the stimulation with a MAMP (microbe-associated molecular pattern), a pathogen, or a secondarily produced damage-associated molecular pattern (DAMPs) and is thus also used as an important bioassays to study receptor–ligand systems related to plant defense [7]. However, since the generation of ROS can be measured during developmental processes such as flower abscission as well, the assay is also suitable to study receptor–ligand interactions of systems not related to plant defense and stress cognition [2].

Extracellular ROS are generated when electrons are transferred from cytosolic NADPH to apoplastic oxygen to generate superoxide (O_2^-) [8]. This step is catalyzed by NADPH oxidases, while a special role during plant immune and stress responses could be assigned to RBOHD [9]. The produced O_2^- is mainly converted to hydrogen peroxide (H_2O_2) by superoxide dismutases [3, 10], and the free H_2O_2 is used in the plant cell wall to cross-link constituents leading to a reinforcement that prevents the plant from further attacks. Besides, the produced ROS can also directly affect the microbes and the pathogen growth.

Since the 1980s, a few bioassays such as the staining with 3,3′-diaminobenzidine (DAB) [11, 12] have been established to monitor H_2O_2 or ROS production. Here, we want to focus on a luminol-based assay using the luminol variant L-012 (8-Amino-5-chloro-7-phenylpyrido [3,4-d] pyridazine-1,4-(2H,3H) dione sodium salt; http://www.wako-chem.co.jp/english/labchem/journals/wpu_biol/10.htm) as a substrate [13]. The substrate is oxidized by peroxidases in the presence of ROS under the emission of light that can be detected by a luminometer or a sensitive CCD camera system. This assay allows for continuous, "online," monitoring of ROS synthesis in vivo and analysis of the kinetics of this cellular response. The original luminol-based assay was described by Keppler et al. [14] and subsequently adapted for other plant cells and tissues [15, 16] and also used in recent publications [2, 17, 18].

2 Materials

2.1 Plant Material and Sample Preparation

1. Water (ddH_2O, has not to be sterile).
2. Petri dishes.
3. Scissors.
4. Plant leaf pieces (any plant of interest, e.g., *Arabidopsis* or *Nicotiana benthamiana* transiently expressing a construct of interest; any binary vector for expression *in planta* is suitable).

2.2 Tools to Set Up the Assay

1. Small spatula.
2. Luminol L-012: 200 μM (Wako Chemicals USA).
3. Horseradish peroxidase: 10 μg/ml (AppliChem) (*see* **Note 1**).
4. Elicitors of interest: synthesized or purified MAMPs, DAMPs, peptides, or crude extracts.
5. Positive control; e.g., well-characterized and known MAMP (e.g., flg22, chitin oligomers, etc.).
6. 10× luminol master mix: 200 μM luminol L-012; 10 μg/ml horseradish peroxidase.
7. 96-well plates, white, flat-bottomed (e.g., LIA plate, white, 96-well, flat-bottom, Greiner Bio-One GmbH), or suitable cuvettes for a single cell instrument.
8. Luminometer, either a 96-well plate reader (e.g., Centro LB 960, Microplate Luminometer, Berthold Technologies) or a single cell instrument.

3 Methods

3.1 Preparation of Plant Leaf Pieces

Before starting the experiment, you might ensure that your receptor protein is properly expressed. Western blot and fluorescent microscopy (in case the receptor of interest is tagged by, e.g., GFP) are well-suiting methods. However, it is also possible to check any non-transformed plant; the ROS assay described here is also a good method on its own to check whether there exists a reactive receptor in a respective plant species or not.

One day before ROS burst measurements, the plant leaves are cut into small pieces:

1. Label petri dishes appropriately and fill them half with deionized water (need not be sterile).
2. Take younger or highly differentiated leaves of nonflowering plants (about 4–6 weeks), and cut them into small, equally sized pieces (e.g., 3 × 3 mm) (*see* **Note 2**) that will fit into the wells of a 6-well plate. At least 3–4 replicates per planned treatment should be prepared (*see* **Note 3**). Do not forget that you need leaf pieces also for negative and positive controls.
3. Place the leaf pieces floating on the water surface in petri dishes. While orientation of pieces (upside up or down) seems not to matter, submersion should be avoided!
4. Cover the leaf samples in the petri dish with a lid and incubated overnight (or >8 h) at room temperature without shaking (*see* **Note 4**).

3.2 Preparation of Samples Prior to Luminescence Measurements, Next Day

1. For a standard 96-well plate-reading luminometer, prepare 90 μl water +10 μl luminol master mix (10×), and pipette into each well of a new 96-well plate (100 μl per well). For other luminometers (e.g., single cell), this might be scaled up 2–3 times, according to the size of the measurement cuvette.
2. Add one leaf piece per well; or if they are small, use two per well. Mechanical damage and wounding must be avoided since this can disturb the measurements by increasing the background! Do not use forceps but only a small spatula! Check if all leaf pieces properly swim on the luminol mix in the center of the well before starting the measurement.

3.3 Luminometer Measurements

Depending on individual requirements, the luminometer settings can be adjusted as necessary (*see* **Note 5**). As an example, in our last publications, we used the following setup for a 96-well reader: Recording time per measurement and well =1 s (could be shorter, depending on the limits of your machine and individual requirements); the time per cycle might be adjusted according to the number of total samples and the duration of recording time per well. Measurement cycles (30–60 s, each) can be repeated to a total time of 20–60 min or longer if necessary.

1. Measure the background level before the induction of a ROS burst with MAMPs/elicitors, to ensure constant values of light emission over time without any stimulus. A measurement for about 5–10 min is usually enough to see a stable baseline. The background measurement might be prolonged up to 60 min if there is no stable baseline within the first 10 min.
2. Start the assay by the addition of the MAMP or any other triggering substance (usually 1 μl of a 1–100 μM stock solution into 100 μl; resulting in a 1:100 dilution with a final concentration of 10–1000 nM). Be aware that the detection of ROS is a pH-sensitive reaction (*see* Fig. 1, **Notes 6** and **7**). Shake the plate shortly (e.g., horizontally on a table to avoid spillovers) to ensure good mixing of triggering substances, and immediately put back to the machine for the measurement. Remember negative (e.g., mock treated with 0.01 mg/ml BSA) and positive controls (e.g., 100 nM of flg22, the peptide epitope of bacterial flagellin that triggers ROS production in nearly all plants).
3. Process the data by determination and calculation of the half maximal effective concentration (EC50) values (*see* **Note 8**). In general, differences in the ox-burst kinetics might be observed depending on the trigger or its corresponding receptor (*see* **Note 9**). Be aware that the optimal conditions for ligand binding/receptor activation and the detection of the ROS burst by chemiluminescence sometimes run in the opposite directions (*see* **Note 10**).

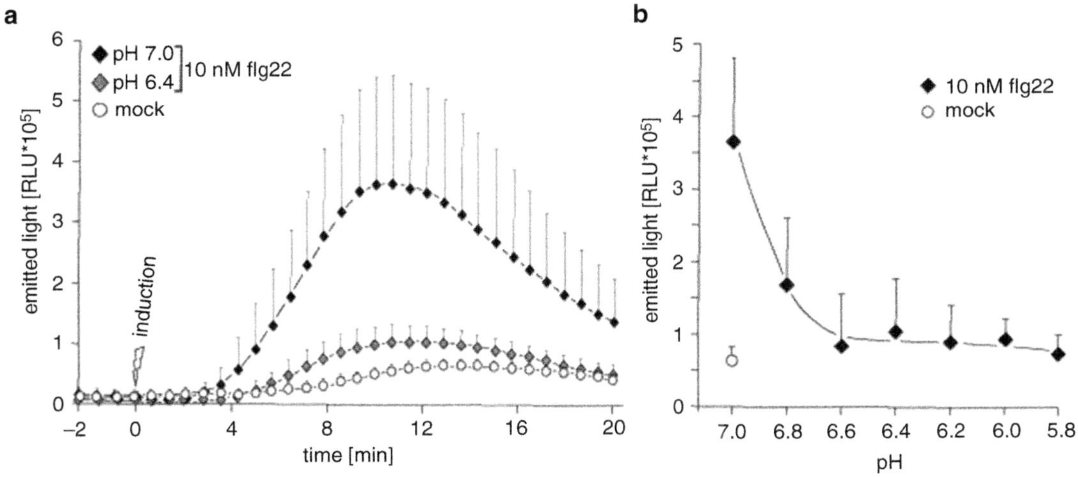

Fig. 1 The influence of pH on the detection of ROS burst in the chemiluminescent bioassay. Leaf samples of *A. thaliana* (Col-0) were equally treated with 10 nM flg22 as an inducer of ROS burst in solutions buffered with 2 mM MES at pH values indicated; mock = 0.01 mg/ml BSA. (**a**) ROS burst was detected as emitted light over time in the L-012-based assay by a 96-well luminometer (Centro LB960, Berthold Life Technologies). (**b**) pH-dependent drop of chemiluminescence; diagram shows maxima of kinetic curves plotted against pH. (**a** and **b**) Values represent means and standard deviations of $n = 4$ technical replicates

4 Notes

1. Be aware that many plant tissues might contain sufficient endogenous peroxidase activity.
2. The condition of the leaf pieces is important for measuring the ROS production kinetics. Spontaneous induction of ROS in the absence of an additional stimulus may arise from the handling of the leaf pieces or from (pre-existing) microbial infections of the leaf pieces.
3. For experiments it is important to use negative and positive controls, and leaf pieces should be equal in size. Technical replicates ($n \geq 3$) are necessary to perform statistics, and independent repetitions of the experiments are required to substantiate any hypothesis.
4. Slow and weak ROS responses may result from suboptimal periods between cutting of the leaves and measuring (e.g., <6 h or >18 h).
5. Chemiluminescence is given in RLU (relative light units) when measured by a luminometer. Since RLU is not a clearly defined unit, it can vary from machine to machine. Thus, it might be problematic to compare the results obtained with different machines.

6. The oxidation of the L-012 substrate and thus the detection of ROS are a pH-sensitive reaction with an optimum at pH ~7–8. (The usage instructions for L-012 suggest a pH optimum at 7.5; Wako Pure Chemical Industries, Ltd.) Hence, it is very important to not use buffers or reagents that lower (or increase) the pH too much. An example is given in Fig. 1, where 2 mM MES buffer influences the ox-burst by drastically lowering the amplitude even when used at pH of 6.6. Therefore, even adding only 1 μl of a test substance buffered in 25 mM MES (at pH 5–6) to the assay solution of 100 μl could lead to a decreased, in worst case, totally abolished light reaction and non-detectable ROS production.
7. Some reagents or contents might directly influence the luminol oxidation, independent from pH values. For example, crude, non-purified extracts of microbes or plants often contain light-quenching constituents (phenolic substances or other secondary metabolites). Besides quenching effects, degrading enzymes can be present in crude extracts (e.g., proteases, hydrolases) that destroy bioactive molecules. If there is no increased chemiluminescence observed within 1 h, quenching effects can be tested by subsequent addition of a functional trigger (e.g., the MAMP flg22).
8. For data handling and evaluation, please refer to [16]. Therein it is in detail described how to measure a dose-response curve and how to determine the EC50. This evaluation tests the sensitivity of a biological system for a given type of trigger.
9. One notable alteration is noticed when comparing MAMP-triggered ox-burst either induced via receptor kinases (RKs) or by receptor-like proteins (RLPs). While, e.g., the flg22-triggered ROS production induced by the RK FLS2 (flagellin-sensitive 2) occurs faster and with higher amplitude, the RLP-triggered ox-burst via Cuscuta receptor 1 (CuRe1) [19] has a much longer lag phase, and the maximum response is clearly lower (Fig. 2). This delayed and lower ROS response was reproducibly seen for this RLP-mediated ox-burst.
10. Notably, many receptors detect extracellular signals at apoplastic pH values at 5–6. For example, the *Arabidopsis* RLP23 detects its ligand peptide nlp20 and induces the production of ethylene best at pH ~5.7 (MES buffered) [17].

Acknowledgments

M.A. was supported by the "Deutsche Forschungsgemeinschaft" (AL1426/1-2), and U.F. was also supported by the German Federal Ministry of Education (BMBF–KBBE project 031A328 36). We thank Georg Felix for critical reading of the manuscript.

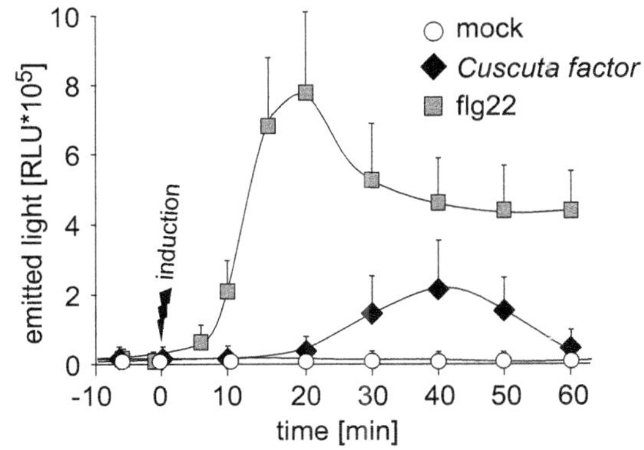

Fig. 2 Comparison of peptide-triggered ROS burst by two different types of receptors in tomato. Tomato leaf pieces were either triggered with a peptide of the parasitic plant *Cuscuta reflexa* (Cuscuta factor) or the bacterial MAMP flg22. While flg22 is perceived by the receptor kinase FLS2, the Cuscuta factor is perceived by the receptor-like protein Cuscuta receptor 1 (CuRe1) [19]. The detected ROS burst over time clearly shows differences in the length of the lag phases (time between induction and raise of the oxidative burst) and in the height of the maximum response. Values represent means and standard deviations of $n = 4$ technical replicates

References

1. Tripathy BC, Oelmuller R (2012) Reactive oxygen species generation and signaling in plants. Plant Signal Behav 7(12):1621–1633
2. Butenko MA, Wildhagen M, Albert M, Jehle A, Kalbacher H, Aalen RB, Felix G (2014) Tools and strategies to match peptide-ligand receptor pairs. Plant Cell 26(5):1838–1847
3. Suzuki N, Koussevitzky S, Mittler R, Miller G (2012) ROS and redox signalling in the response of plants to abiotic stress. Plant Cell Environ 35(2):259–270
4. Boller T, Felix G (2009) A renaissance of elicitors: perception of microbe-associated molecular patterns and danger signals by pattern-recognition receptors. Annu Rev Plant Biol 60:379–406
5. Torres MA, Jones JD, Dangl JL (2006) Reactive oxygen species signaling in response to pathogens. Plant Physiol 141(2):373–378
6. Doke N (1983) Generation of superoxide anion by potato-tuber protoplasts during the hypersensitive response to hyphal wall components of phytophthora-infestans and specific-inhibition of the reaction by suppressors of hypersensitivity. Physiol Plant Pathol 23(3):359–367
7. Dixon RA, Harrison MJ, Lamb CJ (1994) Early events in the activation of plant defense responses. Annu Rev Phytopathol 32:479–501
8. Lamb C, Dixon RA (1997) The oxidative burst in plant disease resistance. Annu Rev Plant Physiol Plant Mol Biol 48:251–275
9. Kadota Y, Shirasu K, Zipfel C (2015) Regulation of the NADPH oxidase RBOHD during plant immunity. Plant Cell Physiol 56(8):1472–1480
10. Marino D, Dunand C, Puppo A, Pauly N (2012) A burst of plant NADPH oxidases. Trends Plant Sci 17(1):9–15
11. Torres MA, Jones JD, Dangl JL (2005) Pathogen-induced, NADPH oxidase-derived reactive oxygen intermediates suppress spread of cell death in *Arabidopsis thaliana*. Nat Genet 37(10):1130–1134
12. ThordalChristensen H, Zhang ZG, Wei YD, Collinge DB (1997) Subcellular localization of H_2O_2 in plants. H_2O_2 accumulation in papillae

and hypersensitive response during the barley-powdery mildew interaction. Plant J 11(6): 1187–1194

13. Teranishi K, Nishiguchi T (2004) Cyclodextrin-bound 6-(4-methoxyphenyl)imidazo[1,2-alpha+/−]pyrazin-3(7H)-ones with fluorescein as green chemiluminescent probes for superoxide anions. Anal Biochem 325(2):185–195
14. Keppler LD, Baker CJ, Atkinson MM (1989) Active oxygen production during a bacteria-induced hypersensitive reaction in tobacco suspension cells. Phytopathology 79:974–978
15. Felix G, Duran JD, Volko S, Boller T (1999) Plants have a sensitive perception system for the most conserved domain of bacterial flagellin. Plant J 18:265–276
16. Albert M, Butenko MA, Aalen RB, Felix G, Wildhagen M (2015) Chemiluminescence detection of the oxidative burst in plant leaf pieces. Bioprotocolorg 5(6):e1423. doi:10.21769/BioProtoc.1423
17. Albert I, Bohm H, Albert M, Feiler CE, Imkampe J, Wallmeroth N, Brancato C, Raaymakers TM, Oome S, Zhang H, Krol E, Grefen C, Gust AA, Chai J, Hedrich R, Van den Ackerveken G, Nurnberger T (2015) An RLP23-SOBIR1-BAK1 complex mediates NLP-triggered immunity. Nature Plants 1:15140
18. Albert M, Jehle AK, Furst U, Chinchilla D, Boller T, Felix G (2013) A two-hybrid-receptor assay demonstrates heteromer formation as switch-on for plant immune receptors. Plant Physiol 163(4):1504–1509
19. Hegenauer V, Fürst U, Kaiser B, Smoker M, Zipfel C, Felix G, Stahl M, Albert M (2016) Detection of the plant parasite *Cuscuta reflexa* by a tomato cell surface receptor. Science 353(6298):478–481

Part III

In silico Methods for Identification and Analyses of PRKs

Chapter 8

Extraction and Curation of Gene Models for Plant Receptor Kinases for Phylogenetic Analysis

Aleksia Vaattovaara, Jarkko Salojärvi, and Michael Wrzaczek

Abstract

Analysis of gene families and identification of homologous genes are important for phylogenetic analysis and for translating results from model to crop species. While numerous plant genomes have been sequenced and made available, the identification of gene models can be difficult, in particular for large gene families arranged in tandem repeats or encoding proteins with a variable number of internal repeats. Thus, correct annotation of plant receptor kinases (PRK) is a challenge. Here, we describe a workflow for the semi-manual extraction, annotation, and verification of genes from annotated gene models as well as from non-annotated DNA regions. This protocol allows the efficient identification of gene family member of PRK from most available plant genomes.

Key words Annotation, Gene family, Phylogeny, Evolution, Genome, HMMER, Domain, Motif

1 Introduction

The dramatic progress of molecular biology over the last decades is largely due to the concentrated research on a selected set of model species. In plants, the adoption of thale-cress, *Arabidopsis thaliana*, and the subsequent development of numerous molecular tools for the species have allowed unprecedented insights into the biochemical and physiological mechanisms that regulate plant life [1]. Likewise, the availability of well-annotated genomes has had a strong impact on the analysis of genes and their evolution.

The rapid development of high-throughput sequencing techniques has made it possible to assemble bacterial, fungal, animal, and plant genomes with ever-increasing quality. This "genomic revolution" has made it possible to investigate the evolution and conservation of molecular components across different species, and it forms a critical step when transferring knowledge from model to species that have more practical use.

A major obstacle in identifying homologous genes and their subsequent evolutionary analysis is the quality of the annotations

Reidunn Birgitta Aalen (ed.), *Plant Receptor Kinases: Methods and Protocols*, Methods in Molecular Biology, vol. 1621, DOI 10.1007/978-1-4939-7063-6_8, © Springer Science+Business Media LLC 2017

in the available genomes. In particular, large gene families which are expanding through tandem repeats create significant challenges for automatic gene prediction software [2–4]. Thus, annotations which are exclusively created by automated methods can contain mis-annotated genes, pseudogenes, or genes that are missed completely during the annotation process. In the recent silver birch genome project [5], this was benchmarked: 3190 automatically predicted genes were manually annotated, and 27% required manual correction. Manual annotation will drastically improve the quality of gene models. After using manually annotated genes to retrain the gene predictors, the accuracy of automated annotations rose to 92% [5]. Wrong annotations can include partial annotations with missing domains, wrong prediction of exon/intron borders, or fusion of two or more separate genes into a single gene model. Alternatively, a single gene might be split into two separate gene models. Using such low-quality datasets will naturally cause problems in sequence alignments. Subsequently, this can result in low-quality phylogenetic trees. Genes that have been missed during annotation may result in false homologies between species and affect evolutionary inference since sub-, neo-, or nonfunctionalization of genes is likely to occur in case of gene duplication events [6].

While the problems and challenges outlined in the previous paragraph apply to most, if not all, gene families, annotation and extraction of gene family members for subsequent phylogenetic analyses are even more demanding for highly expanded gene families like the plant receptor-like kinases (RLKs). Plant genomes typically encode a very large number of RLKs; the *Arabidopsis* and rice genomes contain approximately 600 and 1100 genes, respectively [7–9]. RLKs in plants have been expanded by full genome duplication events but also by tandem duplications [10]. The large number of RLKs and their different expansions likely reflect specific adaptations of various plant lineages to very diverse environmental challenges and lifestyles. In addition to their large number, the organization of plant RLKs in the genomes can add further difficulties [7, 10]. For example, the cysteine-rich receptor-like kinases (CRKs) are, with few exceptions, arranged in repeats [11, 12], and other RLK families display similar organization [10]. Also, the very structure of RLKs with an extracellular region, a transmembrane domain, and an extremely conserved intracellular kinase domain adds to the difficulty of correctly identifying gene models. The presence of a variable number of leucine-rich repeats (LRRs) in the extracellular, ligand-binding domain of many RLKs further complicates annotation. In summary, while careful annotation and curation of gene models are necessary for all genes and gene families, the RLK family poses some unique challenges due to their protein structure, their sheer number, and their genomic organization.

In this protocol, we describe a simple but efficient workflow for semi-manual identification and curation of genes from annotated and non-annotated genomic sequence data. The set of available predicted gene models is first used to identify genes of interest from predicted protein-coding regions through searches using HMMER [13] and in parallel from DNA sequence using Wise2 [14–16]. The results are validated by using multiple sequence alignments (MSAs) and phylogenetic trees and by verifying the presence of conserved domains and subsequent manual correction of gene models.

2 Materials

2.1 Choosing the Relevant Species for Analyses

The first critical step in the beginning of any evolutionary-based analysis is the selection of relevant species. Plenty of algal and plant genomes of varying quality are available from a number of different databases. Typically, the genomes of model species, including *Arabidopsis* and rice, feature superior genome and gene annotation quality which makes them a good starting point.

Selection of species will depend on the specific aims of the study. Generally, select a good coverage of species from the specific taxonomic group (e.g., Brassicaceae, Eudicots, whole plant kingdom) of interest. Gene families are expanded mainly by two different processes: whole genome duplication events or tandem duplication; this is why it may be interesting to select species that track the history of expansion events. One resource for whole genome duplication events in plants is the Plant Genome Duplication Database [17], whereas tandem duplications can be identified, for example, with CoGe platform [18]. Finally, add one or several species that are suitable as out-group and which make sense in terms of evolution. Out-group species do not necessarily have to be plants if the gene family of interest is not plant specific. A larger number species are frequently favored for broader analyses. However, more species also result in more work in detecting and curating gene models, in particular for gene families with many members.

2.2 Collecting the Data from Databases

The majority of plant genomes are stored in databases (e.g., Phytozome [19], Ensembl Plants [20], Gramene [21], and National Center for Biotechnology Information (NCBI) [22]), while others are available through species-specific webpages or databases like The Arabidopsis Information Resource (TAIR) [23]. Different databases may contain different versions of the genomes, and it is important to check for the newest version available.

Usually, for each genome, the genomic and annotated protein sequences are provided in a FASTA file format. To facilitate the analysis of annotated, mostly predicted, protein sequences, preferably choose files containing only the representative gene models (in Phytozome, this is referred to as "primary transcript only"), as sometimes there are several isoforms/splice variants annotated for one gene. If these are not available, one practical way of choosing the representative gene model is to use the transcript that produces the longest coding sequence. However, this criterion should be used only as a rule of thumb, since proper definition of the representative gene model is a separate research question. For vertebrate animals, methods have been developed which look at structural and functional protein data and information from cross-species conservation to carry out this task [24].

2.3 Motifs

Gene and protein family members can be identified in several different ways. If the protein of interest contains a domain which has been defined in the Pfam database [25], the domain can be utilized to identify gene models from either amino acid (HMMER) or DNA sequence (Wise2), respectively. Pfam domains are defined using hidden Markov model (HMM), which essentially describes the sequence of different amino acid frequencies encountered in an alignment of conserved region derived from MSA. If no preexisting Pfam domains are available, a new domain can be learned using the "hmmbuild" command in HMMER with an existing MSA as input. Typically, the quality of the domain detection will improve with increasing numbers of input sequences.

2.4 Computational Environment

HMMER, MUSCLE [26, 27], and PRANK [28] can be installed to Unix-like environments (including Linux and Mac OS X) or Windows. However, a Unix-like environment like Linux is required to run the Wise2 program. Alternatively, instead of running programs on a workstation or computing cluster, HMMER, MUSCLE, and webPRANK [29] are also available as web services from the European Bioinformatics Institute (EBI; https://www.ebi.ac.uk/). Some genome databases also provide tools for HMMER and BLAST [30] searches. For example, Phytozome provides the BioMart tool [31, 32], which allows searching for gene models from different species based on different IDs such as Pfam domain numbers.

3 Methods

Here, we describe a protocol to identify gene family members from plant genomes and to create a high-quality dataset for further analyses. Figure 1 shows a schematic overview of the workflow. The examples in this protocol are based on Unix versions of the programs.

3.1 Identification of Proteins of Interest

1. Using **HMMER**, scan the amino acid sequences of the annotated gene models for the Pfam (or self-made) domain and extract the identified sequences into a new FASTA file. HMM

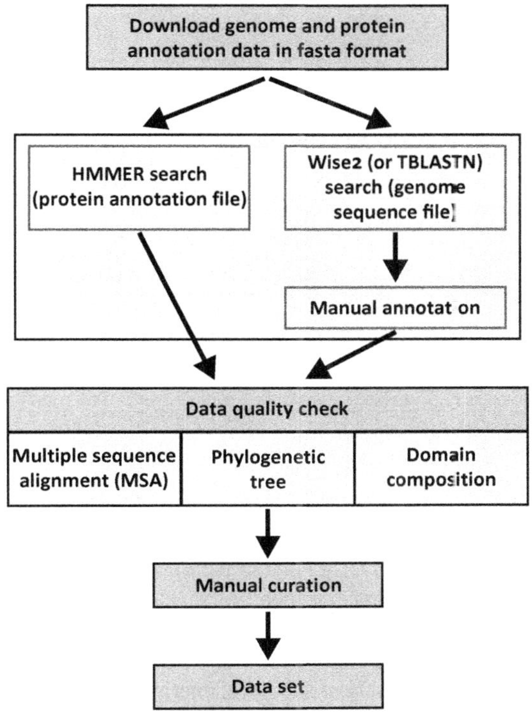

Fig. 1 Workflow for extracting and correcting gene models for subsequent analyses. After downloading the data, search using HMMER and annotated gene models should be carried out in parallel to de novo identification based on DNA sequence using Wise2 and manual annotation. Quality check has to be carried out for all identified gene models

files need to be prepared prior to searching with the "hmmpress" command. An example command for HMMER can be "hmmscan –domtblout output_file_name domain.hmm input_file.fasta." An example output file of "hmmscan" with the "domtblout" option is presented in Fig. 2. The results from columns 13 and 14 are needed to estimate the likelihood for correct domain identification. A high i-Evalue (column 13) and a score (column 14) close to or below zero indicate that the result for a given domain is very likely false. Sequences, for which domains were identified with low i-Evalues and simultaneously high scores, can be extracted from original FASTA file based on the sequence identifiers from column 4 (*see* **Note 1**).

2. **Wise2** can be used to search for the HMM domain from the genome sequence (DNA). Ideally, this will yield similar results as the HMMER search but will possibly identify also non-annotated gene models. In order to use domains in Wise2 searches, the HMM file needs to be converted to HMMER2 format (typically files are in the newer HMMER3 format although the ending for both files is the same .hmm) with the

```
#                                                                          --- full sequence --- -
# target name        accession   tlen query name           accession   qlen   E-value  score  bias
#------------------- ---------- ----- -------------------- ---------- ----- --------- ------ ----- -
Pkinase_Tyr          PF07714.12   259 EFR                  -           1031     7e-42  129.6   0.0
Pkinase_Tyr          PF07714.12   259 FLS2                 -           1173   4.8e-43  133.4   0.0
#
# Program:         hmmscan
# Version:         3.1b2 (February 2015)
# Pipeline mode:   SCAN
# Query file:      EFR_FLS2.fasta
# Target file:     Pkinase_Tyr.hmm
# Option settings: hmmscan --domtblout EFR_FLS2_kinase --cpu 6 Pkinase_Tyr.hmm EFR_FLS2.fasta
# Current dir:     /wrk/vaattova/Book_chapter
# Date:            Wed Jun 22 14:50:08 2016
# [ok]
```

```
------------- this domain -------------   hmm coord   ali coord   env coord
 #  of  c-Evalue  i-Evalue  score  bias  from    to  from    to  from    to  acc description of target
-- ---  --------- --------- ------ ----- ----- ----- ----- ----- ----- ----- ---- ---------------------
 1   1    9.3e-42   9.3e-42  129.2   0.0     4   208   715   933   712   993 0.81 Protein tyrosine kinase
 1   1    1.1e-42   1.1e-42  132.3   0.0     5   259   874  1153   870  1153 0.82 Protein tyrosine kinase
```

Fig. 2 Sample output from HMMER showing example output using the "domtblout" option. The output file is a table listing the domains that have been found from the set of protein models used as input. In this example, the protein sequences of two receptor-like protein kinases EF-Tu RECEPTOR (EFR; AT5G20480) and FLAGELLIN-SENSITIVE 2 (FLS2; AT5G46330) were given as input in a FASTA file and the Pfam domain PF07714.12 as input in a HMM file. The command used to create this output can be found from the bottom of the output file. The first three columns in the table provide basic information about the HMM domain (Target). Column 4 contains the name of the protein model. Columns 13 and 14 (i-Evalue and score for domain) are important for estimating the reliability of domain prediction. Detailed description of all columns can be found in the HMMER user guide (http://hmmer.org/)

command "hmmconvert" (example command: "hmmconvert −2 HMMER3_format.hmm > HMMER2_format.hmm"). An example command to run Wise2 can look as follows: "genewisedb -hmmer domain_in_HMMER2_format.hmm -dnadb input_file.fasta > output_file_name." In the output (Fig. 3), only sequences containing the HMM domain are listed (this will typically be chromosomes or contigs from a genome sequence file, depending on the type of input). Each hit is presented as an alignment together with a score value. Due to the nature of the program, comparing a protein domain against DNA sequence using Wise2 results in more false-positive results compared to HMMER. Thus, caution is warranted in particular

Fig. 3 Sample output from Wise2. Here, a part of Wise2 output file is shown. Part of the output has been removed due to space constraints (indicated by *three dots* [...]). In this example, the genomic sequences of EF-Tu RECEPTOR (EFR; AT5G20480) and FLAGELLIN-SENSITIVE 2 (FLS2; AT5G46330) were used as input, including 20,000 nucleotides upstream and downstream, respectively. The *white arrow* points to a list of the DNA stretches found to contain the Pfam domain PF07714.12. *Black arrows* point the scores of the HMM motif to DNA alignments. In this example, Alignment 1 is likely a false-positive result while Alignment 2 is correct. The end of the file output was also removed due to space constraints

```
Wise2 - database searching mode
Program: genewisedb version: $Name: wise2-4-1 $ released: unreleased
This program is freely distributed under a Gnu Public License.
   See -version for more info on copyright
Bugs and credits to Ewan Birney <birney@sanger.ac.uk>
-----------------------------------------------------

Algorithm type:          GeneWise
Search algorithm used: 623
Implementation:          Single Threaded processor (serial)
Search mode:             Single protein vs genomic db
Protein info from:       Pkinase_Tyr_hmm2.hmm
Dna info from:           EFR_FLS2_dna.fasta
Start/End (protein)      default
Gene Paras:              human.gf
Codon Table:             codon.table
Subs error:              1e-05
Indel error:             1e-05
Model splice?            model
Model codon bias?        flat
Model intron bias?       tied
Null model               syn
Alignment Alg            623L
[Warning: Can't fit histogram to a db smaller than 1,000]

•••

#High Score list
#Protein ID                 DNA Str  ID                       Bits Evalue
--------------------------------------------------------------------------
Protein Pkinase_Tyr         DNA [+] AT5G46330                 126.89
Protein Pkinase_Tyr         DNA [+] AT5G20480                 116.49

#Alignments
--------------------------------------------------------------------------

>Results for Pkinase_Tyr vs AT5G46330 (forward) [0]

genewisedb output
Score 125.07 bits over entire alignment.
This will be different from per-alignment scores. See manual for
details
For computer parsable output, try genewisedb -help or read the manual
Scores as bits over a synchronous coding model

                    Alignment 1 Score -355.72 (Bits)

Pkinase_Tyr      100 LTLEDLLSFALQIAKGMEYLESKKLVHRDLAARNVLLEENKVVKISDFG

                     -------------------------------------------------
AT5G46330          4

•••

                    Alignment 2 Score 121.89 (Bits)

Pkinase_Tyr        1 LELGKKLGEGAFGEVVEGTLKGEKEKKEVKVAVKTL-KEEAEEEEKEEF
                      +  + +G+ +  +V++G+L+       + +AVK+L  +E + E+ + F
                     FNSANIIGSSSLSTVYKGQLE-----DGTVIAVKVLNLKEFSAESDKWF
AT5G46330      22674 taagaaagtaataagtagccg     ggagaggagtacacttggtgatt
                     tagcattgcggtgctaagata     agcttctattataatccacaagt
                     cctccctcaccacagcatgaa     tgagtaaaagtagactaaacggc


Pkinase_Tyr       49 LEEAKIMKKLKHENIVKLLGVCT-KEEPLLIVTEYVEGGDLLKFLRKKE
                      +EAK++++LKH+N+VK+LG+     ++    +V+ ++E+G+L++ ++ +
                     YTEAKTLSQLKHRNLVKILGFAWESGKTKALVLPFMENGNLEDTIHGSA
AT5G46330      22806 taggaataccaccacgaatgtgtgagaaagtgcctagagatggaacgtg
                     acacactgataagattattgtcgaggacactttcttaagataactagcc
                     caataagcaaatacgggcagtggaccagatagtatggtacggcctccta
```

Fig. 3 Refer caption on previous page

for results that are supported by low score values (usually very small or even negative values). An identified domain that does not correspond with any annotated gene models and is supported by a high score value indicates that either the entire gene has been missed by automated annotation or that the automated annotation resulted in a faulty gene model. For these genes, annotation needs to be rerun or carried out manually (read instructions from Subheading 3.3; *see* **Note 2**).

3. Also **TBLASTN** can be utilized to identify homologous genes from genome sequence by using protein model as query sequence. This can be effective with strongly conserved gene families. Unlike the approaches using HMMER or Wise2, BLAST has the disadvantage that a search is carried out only for one specific sequence. It is also very important to remember that a good match based on BLAST does not necessarily mean that sequences are orthologs or, in case of more distant species, not even homologs. This is less of a problem with HMMER searches since HMMER, in contrast to BLAST, does not compare single genes. In addition to TBLASTN and Wise2, also the program Exonerate [33] is able to carry out sequence alignments between protein and DNA.

3.2 Evaluation of Results

Thorough data evaluation and verification are critical in order to obtain reliable results. The quality of gene models is typically estimated based on their similarity to known, previously verified, sequences. Differences in domain composition or length can indicate problems with a new gene model (e.g., missing or additional exons). Annotation quality of gene models can be assessed using several methods:

1. **Alignments**: Carry out amino acid MSAs of gene models found with HMMER and Wise2 with programs such as MUSCLE or PRANK. Large breaks or insertions in few sequences can indicate errors in the gene model (Fig. 4).
2. **Phylogenetic trees**: Estimate a phylogenetic tree based on an MSA of the data. Long branches in the tree usually indicate problematic gene models. It is also advisable to check whether the overall topological placement of the new gene models in the tree corresponds to known phylogenetic relationships between species (in a tree having orthologous genes from several species) or related gene families (Fig. 5).
3. **Domain composition for gene families with conserved domain structures**: Specific protein domain composition (e.g., certain domains are always present in conserved order) can be utilized to check the gene model quality. A suitable program for this task is SMART [34], which is also available as web service (http://smart.embl-heidelberg.de/). Other programs can be used to identify more specific features. For example,

```
Species/ *    **                                                    * *   *   **
1. Seq 1 TK-FAGTLGYTAPEYALHGQLSEKVDTYSFGIVVLEIVSGKKSSEMIADPGAEYLLKKAWKLYENGAHLELV
2. Seq 2 TK-FAGTLGYTAPEYALHGQLSEKVDTYSFGIVVLEIVSGTKSSEMIADPGAEYLLKKAWKLYEDGTHLELV
3. Seq 3 TRRVVGTFGYMAPEYAMGGLFSVKSDVFSFGVILLEITSGKRSSGFYLSEHRQTLLAYAWRLWNEGREMELV
4. Seq 4 TARIVGTYGYMAPEYAMEGLYSTKSDVFSFGVLLLEIITGKKNSEFNKSKRAPSLLAYAWQLWNEGKELELI
5. Seq 5 TARIVGTY-------------------------------------------------IAWQLWNEGKELELI
6. Seq 6 TNRIVGTYGYMAPEYAMEGLSSVKSDVFSFGVLMLEIISGKRNGGFHLSEEGKSLLNFTWKLWSEGKGLELM
7. Seq 7 TNRIVGTYGYMAPEYAMVGLFSVKSDVFSFGVLLLEIISGEKNVGFHLSEEGESLLTFAWKLWSDGQGLELM
```

Fig. 4 Example of an alignment highlighting problems that can be due to incorrect gene models. This alignment shows a part of the kinase domain from seven different gene models. There is likely an exon missing in the gene model referred to as Seq_5. Thus, Seq_5 will require manual correction

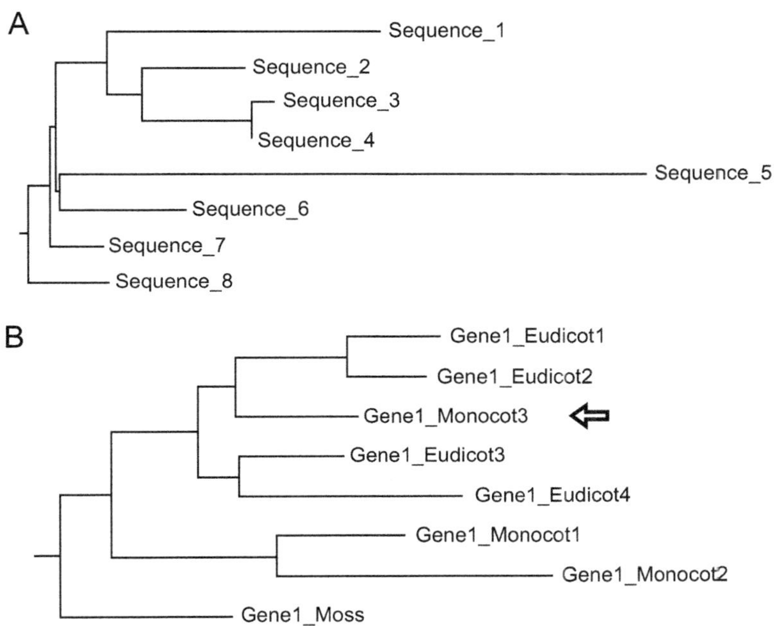

Fig. 5 Example of a phylogenetic tree highlighting problems possibly caused by incorrect gene models. (**a**) The branch length for sequence_5 is very long suggesting that there might be a problem with the gene model requiring manual verification. (**b**) Gene1_Monocot3 resides within a Eudicot-specific cluster. This tree topology is not impossible but unlikely. Thus, the gene model for Gene1_Monocot3 should be manually verified

TMHMM [35, 36] can be used to describe and validate membrane protein topology and transmembrane regions. An important, but frequently overlooked, criterion is to check for complete (or intact) domains. In some cases, even if the gene model is missing an entire exon, programs such as SMART are still able to identify a domain. However, those problems are typically easily visible in MSAs.

3.3 Manual Annotation

If gene models appear incorrect or have not been annotated at all (i.e., the domains are identified from the genome sequence with Wise2 or BLAST), manual curation or annotation is required.

1. Available RNA evidence, for example, expressed sequence tags (ESTs) or RNA sequencing (RNAseq) data, can provide reliable evidence of transcribed sequences. However, it is necessary to verify that the supporting evidence forms a full transcript since next-generation sequencing approaches (RNAseq) are mostly short-read assemblies of transcripts.
2. In the absence of RNA evidence, which is often the case for problematic or missing gene models, domain composition or comparison with related genes or proteins can be used to manually annotate a gene model (*see* **Notes 3** and **4**). Programs such as FGENESH+ [37] accept a stretch of genome sequence (DNA) and available protein model evidence as input and use this information to create a gene model. JBrowse [38] is a convenient tool for extracting genomic DNA sequences which are needed for FGENESH+. The length of the DNA stretch selected for manual annotation depends on the distance to the neighboring genes. In particular, when annotating similar genes in tandem repeats with short intergenic regions between genes, it is critical to ensure that the selected DNA stretch contains a single gene only (information on the distance between gene models can be calculated from Wise2 output for non-annotated genes or alternatively estimated from JBrowse for annotated genes). The quality of the end result depends on the similarity between the genes to be annotated and the protein sequences that are used as models. Sometimes, several attempts using different protein sequences as models can be required in order to obtain an accurate and complete prediction. Particularly for RLKs, the transmembrane region can cause difficulties during annotation.

4 Notes

1. HMMER only identifies gene models which are annotated and present in the FASTA file containing the amino acid sequence of gene models. The problem is that not all genes are annotated at all and many models can be incorrect; e.g., gene model can be partial or a combination from different genes.
2. Some genes may have evolved into pseudogenes during evolution. While these can sometimes still be annotated, they frequently contain STOP codons, resulting in aberrant domain structure of the gene product. Occasionally, domains can be recognized from pseudogenes. This can make manual annotations difficult or even impossible.
3. How to handle partial data? Not all genes can be fully annotated; genes can reside in genome regions with low sequence

coverage or low sequencing quality resulting in a high number of Ns (unknown nucleotides) in a genomic assembly. A gene might also reside at the edge of a contig in a way that makes it impossible to define the beginning or end of the model. While there are some problems with using partial gene models, they can be used in analyses for estimating the number of gene family members in a given species.

4. Genome drafts are not perfect. Similar to gene models, also genome sequences can contain errors. Errors can result from sequencing errors (ambiguous nucleotides) or problems with genome assembly.

Acknowledgments

AV is supported by the Doctoral Programme in Plant Sciences (DPPS) of the University of Helsinki. JS is supported by the University of Helsinki (3-year fund allocation). MW is supported by the Academy of Finland (#275632 and #283139) and the University of Helsinki (3-year fund allocation). The authors are members in the Centre of Excellence (CoE) in the Molecular Biology of Primary Producers (2014–2019) funded by the Academy of Finland (decision #271832).

References

1. Allahverdiyeva Y, Battchikova N, Brosché M, Fujii H, Kangasjärvi S, Mulo P, Mähönen AP, Nieminen K, Overmyer K, Salojärvi J, Wrzaczek M (2015) Integration of photosynthesis, development and stress as an opportunity for plant biology. New Phytol 208(3): 647–655
2. Stanke M, Keller O, Gunduz I, Hayes A, Waack S, Morgenstern B (2006) AUGUSTUS: ab initio prediction of alternative transcripts. Nucleic Acids Res 34(Web Server issue):W435–W439. gkl200
3. Stanke M, Schoffmann O, Morgenstern B, Waack S (2006) Gene prediction in eukaryotes with a generalized hidden Markov model that uses hints from external sources. BMC Bioinformatics 7:62
4. Stanke M, Tzvetkova A, Morgenstern B (2006) AUGUSTUS at EGASP: using EST, protein and genomic alignments for improved gene prediction in the human genome. Genome Biol 7(Suppl 1):S11 11–S11 18
5. Salojärvi J, Smolander OP, Nieminen K , Rajaman S, Mohammadi O, Safdari P, Lamminmäki A, Immanen J, Lan T, Tanskanen J, Rastas P, Amiryousefi A, Jayaprakash B, Kammonen J, Hagqvist R, Eswaran G, Ahonen VH, Serra JA, Asiegbu FO, de Dios Barajas-Lopez J, Blande D, Blokhina O, Blomster T, Broholm S, Brosché M, Cui F, Dardick C, Ehonen E, Elomaa P, Escamez S, Fagerstedt KV, Fujii H, Gauthier A, Gollan PJ, Halimaa P, Heino P, Himanen K, Hollender C, Kangasjärvi S, Kauppinen L, Kelleher CT, Kontunen-Soppela S, Koskinen JP, Kovalchuk A, Kärenlampi S, Kärkönen A, Lim KJ, Leppälä J, Macpherson L, Mikola J, Mouhu K, Mähönen AP, Niinemets Ü, Oksanen E, Overmyer K, Palva ET, Pazouki L, PennanenV, Puhakainen T, Poczai P, Possen BJHM, Punkkinen M, Rahikainen M, Rousi M, Ruonala R, van der Schoot C, Shapiguzov A, Sierla M, Sipilä TP, Sutela S, Teeri TH, Tervahauta AI, Vaattovaara A, Vahala J, Vetchinnikova L, Welling A, Wrzaczek M, Xu E, Paulin L, Schulman AH, Lascoux M, Albert VA, Auvinen P, Helariutta Y, Kangasjärvi K. 2017. Adaptive landscape and population genomics of silver birch. Nat Genet (accepted for publication).
6. Li Z, Defoort J, Tasdighian S, Maere S, Van de Peer Y, De Smet R (2016) Gene duplicability of core genes is highly consistent across all angiosperms. Plant Cell 28(2):326–344

7. Shiu SH, Bleecker AB (2003) Expansion of the receptor-like kinase/Pelle gene family and receptor-like proteins in Arabidopsis. Plant Physiol 132(2):530–543
8. Shiu SH, Bleecker AB (2001) Plant receptor-like kinase gene family: diversity, function, and signaling. Sci STKE 113:re22
9. Shiu SH, Bleecker AB (2001) Receptor-like kinases from arabidopsis form a monophyletic gene family related to animal receptor kinases. Proc Natl Acad Sci U S A 98(19):10763–10768
10. Hofberger JA, Nsibo DL, Govers F, Bouwmeester K, Schranz ME (2015) A complex interplay of tandem- and whole-genome duplication drives expansion of the L-type lectin receptor kinase gene family in the brassicaceae. Genome Biol Evol 7(3):720–734
11. Bourdais G, Burdiak P, Gauthier A, Nitsch L, Salojärvi J, Rayapuram C, Idänheimo N, Hunter K, Kimura S, Merilo E, Vaattovaara A, Oracz K, Kaufholdt D, Pallon A, Anggoro DT, Glow D, Lowe J, Zhou J, Mohammadi O, Puukko T, Albert A, Lang H, Ernst D, Kollist H, Brosché M, Durner J, Borst JW, Collinge DB, Karpinski S, Lyngkjaer MF, Robatzek S, Wrzaczek M, Kangasjäarvi J, Consortium CRK (2015) Large-scale phenomics identifies primary and fine-tuning roles for crks in responses related to oxidative stress. PLoS Genet 11(7):e1005373
12. Wrzaczek M, Brosché M, Salojärvi J, Kangasjärvi S, Idänheimo N, Mersmann S, Robatzek S, Karpinski S, Karpinska B, Kangasjärvi J (2010) Transcriptional regulation of the CRK/DUF26 group of receptor-like protein kinases by ozone and plant hormones in arabidopsis. BMC Plant Biol 10:95
13. Eddy SR (1998) Profile hidden Markov models. Bioinformatics 14(9):755–763
14. Birney E, Durbin R (2000) Using GeneWise in the Drosophila annotation experiment. Genome Res 10(4):547–548
15. Li W, Cowley A, Uludag M, Gur T, McWilliam H, Squizzato S, Park YM, Buso N, Lopez R (2015) The EMBL-EBI bioinformatics web and programmatic tools framework. Nucleic Acids Res 43(W1):W580–W584
16. McWilliam H, Li W, Uludag M, Squizzato S, Park YM, Buso N, Cowley AP, Lopez R (2013) Analysis tool web services from the EMBL-EBI. Nucleic Acids Res 41(Web Server issue):W597–W600
17. Lee TH, Tang H, Wang X, Paterson AH (2013) PGDD: a database of gene and genome duplication in plants. Nucleic Acids Res 41(Database issue):D1152–D1158
18. Lyons E, Pedersen B, Kane J, Alam M, Ming R, Tang H, Wang X, Bowers J, Paterson A, Lisch D, Freeling M (2008) Finding and comparing syntenic regions among arabidopsis and the outgroups papaya, poplar, and grape: CoGe with rosids. Plant Physiol 148(4): 1772–1781
19. Goodstein DM, Shu S, Howson R, Neupane R, Hayes RD, Fazo J, Mitros T, Dirks W, Hellsten U, Putnam N, Rokhsar DS (2012) Phytozome: a comparative platform for green plant genomics. Nucleic Acids Res 40(Database issue):D1178–D1186
20. Kersey PJ, Allen JE, Armean I, Boddu S, Bolt BJ, Carvalho-Silva D, Christensen M, Davis P, Falin LJ, Grabmueller C, Humphrey J, Kerhornou A, Khobova J, Aranganathan NK, Langridge N, Lowy E, McDowall MD, Maheswari U, Nuhn M, Ong CK, Overduin B, Paulini M, Pedro H, Perry E, Spudich G, Tapanari E, Walts B, Williams G, Tello-Ruiz M, Stein J, Wei S, Ware D, Bolser DM, Howe KL, Kulesha E, Lawson D, Maslen G, Staines DM (2016) Ensembl genomes 2016: more genomes, more complexity. Nucleic Acids Res 44(D1):D574–D580
21. Tello-Ruiz MK, Stein J, Wei S, Preece J, Olson A, Naithani S, Amarasinghe V, Dharmawardhana P, Jiao Y, Mulvaney J, Kumari S, Chougule K, Elser J, Wang B, Thomason J, Bolser DM, Kerhornou A, Walts B, Fonseca NA, Huerta L, Keays M, Tang YA, Parkinson H, Fabregat A, McKay S, Weiser J, D'Eustachio P, Stein L, Petryszak R, Kersey PJ, Jaiswal P, Ware D (2016) Gramene 2016: comparative plant genomics and pathway resources. Nucleic Acids Res 44(D1):D1133–D1140
22. Clark K, Karsch-Mizrachi I, Lipman DJ, Ostell J, Sayers EW (2016) GenBank. Nucleic Acids Res 44(D1):D67–D72
23. Berardini TZ, Reiser L, Li D, Mezheritsky Y, Muller R, Strait E, Huala E (2015) The arabidopsis information resource: making and mining the "gold standard" annotated reference plant genome. Genesis 53(8): 474–485
24. Rodriguez JM, Carro A, Valencia A, Tress ML (2015) APPRIS webserver and webservices. Nucleic Acids Res 43(W1):W455–W459
25. Finn RD, Coggill P, Eberhardt RY, Eddy SR, Mistry J, Mitchell AL, Potter SC, Punta M, Qureshi M, Sangrador-Vegas A, Salazar GA, Tate J, Bateman A (2016) The Pfam protein families database: towards a more sustainable future. Nucleic Acids Res 44(D1): D279–D285
26. Edgar RC (2004) MUSCLE: multiple sequence alignment with high accuracy and high throughput. Nucleic Acids Res 32(5): 1792–1797

27. Edgar RC (2004) MUSCLE: a multiple sequence alignment method with reduced time and space complexity. BMC Bioinformatics 5:113
28. Löytynoja A, Goldman N (2008) Phylogeny-aware gap placement prevents errors in sequence alignment and evolutionary analysis. Science 320(5883):1632–1635
29. Löytynoja A, Goldman N (2010) webPRANK: a phylogeny-aware multiple sequence aligner with interactive alignment browser. BMC Bioinformatics 11:579
30. Altschul SF, Gish W, Miller W, Myers EW, Lipman DJ (1990) Basic local alignment search tool. J Mol Biol 215(3):403–410
31. Smedley D, Haider S, Ballester B, Holland R, London D, Thorisson G, Kasprzyk A (2009) BioMart–biological queries made easy. BMC Genomics 10:22
32. Smedley D, Haider S, Durinck S, Pandini L, Provero P, Allen J, Arnaiz O, Awedh MH, Baldock R, Barbiera G, Bardou P, Beck T, Blake A, Bonierbale M, Brookes AJ, Bucci G, Buetti I, Burge S, Cabau C, Carlson JW, Chelala C, Chrysostomou C, Cittaro D, Collin O, Cordova R, Cutts RJ, Dassi E, Di Genova A, Djari A, Esposito A, Estrella H, Eyras E, Fernandez-Banet J, Forbes S, Free RC, Fujisawa T, Gadaleta E, Garcia-Manteiga JM, Goodstein D, Gray K, Guerra-Assuncao JA, Haggarty B, Han DJ, Han BW, Harris T, Harshbarger J, Hastings RK, Hayes RD, Hoede C, Hu S, Hu ZL, Hutchins L, Kan Z, Kawaji H, Keliet A, Kerhornou A, Kim S, Kinsella R, Klopp C, Kong L, Lawson D, Lazarevic D, Lee JH, Letellier T, Li CY, Lio P, Liu CJ, Luo J, Maass A, Mariette J, Maurel T, Merella S, Mohamed AM, Moreews F, Nabihoudine I, Ndegwa N, Noirot C, Perez-Llamas C, Primig M, Quattrone A, Quesneville H, Rambaldi D, Reecy J, Riba M, Rosanoff S, Saddiq AA, Salas E, Sallou O, Shepherd R, Simon R, Sperling L, Spooner W, Staines DM, Steinbach D, Stone K, Stupka E, Teague JW, Dayem Ullah AZ, Wang J, Ware D, Wong-Erasmus M, Youens-Clark K, Zadissa A, Zhang SJ, Kasprzyk A (2015) The BioMart community portal: an innovative alternative to large, centralized data repositories. Nucleic Acids Res 43(W1):W589–W598
33. Slater GS, Birney E (2005) Automated generation of heuristics for biological sequence comparison. BMC Bioinformatics 6:31
34. Schultz J, Milpetz F, Bork P, Ponting CP (1998) SMART, a simple modular architecture research tool: identification of signaling domains. Proc Natl Acad Sci U S A 95(11): 5857–5864
35. Letunic I, Doerks T, Bork P (2015) SMART: recent updates, new developments and status in 2015. Nucleic Acids Res 43(Database issue):D257–D260
36. Krogh A, Larsson B, von Heijne G, Sonnhammer EL (2001) Predicting transmembrane protein topology with a hidden Markov model: application to complete genomes. J Mol Biol 305(3):567–580
37. Solovyev V, Kosarev P, Seledsov I, Vorobyev D (2006) Automatic annotation of eukaryotic genes, pseudogenes and promoters. Genome Biol 7(Suppl 1):S10 11–S10 12
38. Skinner ME, Uzilov AV, Stein LD, Mungall CJ, Holmes IH (2009) JBrowse: a next-generation genome browser. Genome Res 19(9):1630–1638

Chapter 9

In Silico Prediction of Ligand-Binding Sites of Plant Receptor Kinases Using Conservation Mapping

Russell J.S. Orr and Reidunn Birgitta Aalen

Abstract

Plasma membrane-bound plant receptor-like kinases (RLKs) can be categorized based on their ligand-binding extracellular domain. The largest group encompasses RLKs having ectodomains with leucine-rich repeats (LRRs). The LRR-RLKs can further be assigned to classes mainly based on the number of LRRs. Many of the receptors of the classes X and XI with more than 20 LRRs are activated by small secreted peptide ligands. To understand how peptide signaling works, it is of interest to identify the amino acids of the receptor that are directly involved in ligand interaction. Such residues have most likely been conserved over evolutionary time and can therefore be predicted to be conserved in receptor orthologues of different plant species. Here we present an in silico method to identify such residues. This involves a simplified method for identification of orthologues and a web-based program for identifying the most conserved amino acids aside from the leucines that structure the ectodomain. The method has been validated for the LRR-RLKs HAESA (HAE) and PHYTOSULFOKINE RECEPTOR1 (PSKR1) for which conservation-mapping results closely matched recent structure-based identification of ligand and co-receptor-interacting residues.

Key words Phylogenetic analyses, Leucine-rich repeat receptor-like kinases, Conserved ligand-binding amino acids, Peptide signaling, Ligand-receptor interaction

1 Introduction

Over the last decade, cell-to-cell communication during plant growth and development as well as pathogen defense has been shown to be mediated by small peptides [1–3]. Signaling peptides have been discovered using genetic screens, bioinformatic analyses, and biochemical isolation of molecules affecting growth and development. In the *Arabidopsis thaliana* genome, for example, there are about 1000 putative genes encoding potentially secreted signaling peptides [4]. There are two major groups of secreted peptides—the cysteine-rich, with typically about 50 amino acids that attain their three-dimensional structure by disulfate bridges between pairs of cysteine residues in conserved positions, and 5–30 amino acid long peptides derived from the

Reidunn Birgitta Aalen (ed.), *Plant Receptor Kinases: Methods and Protocols*, Methods in Molecular Biology, vol. 1621,
DOI 10.1007/978-1-4939-7063-6_9,

C-termini of propeptides. Underway, the propeptides may be processed and posttranslationally modified by sulfation of tyrosines, hydroxylation of prolines, and glycosylation of hydroxyprolines (reviewed in [1]).

Secreted signaling peptides are expected to interact with membrane-bound receptor kinases, but at present only a few percent of the potential peptides have been matched with their receptor [3, 5–7]. In *Arabidopsis*, about 600 receptor-like kinases (RLKs) have been identified. They can be assigned to different structural classes based on the extracellular domain (ectodomain) [8]. Thirteen classes with altogether more than 200 members have ectodomains with leucine-rich repeats (LRRs). Interestingly, all the RLKs that have been shown to bind small posttranslationally modified signaling peptides belong to the classes X and XI and have ectodomains with more than 20 LRRs. Genetic evidence, in vitro binding assays, and studies of phenotypes resulting from overexpression or use of synthetic peptides suggest that similar receptors bind ligands from the same peptide family and often influence the same biological processes [5, 9–18]. A high concentration and/or ectopic presence of a signaling peptide may, however, result in activation of non-native receptors. To evaluate whether related LRR-RLKs are likely to interact with members of given peptide gene families, it is valuable to know the residues involved in peptide ligand binding. Ultimately, this can be resolved by determination of the 3D structure of the ectodomain in the presence of the ligand. As long as structure data is lacking, in silico prediction of interacting amino acids can be useful.

Phylogenetic analyses have suggested that some signaling peptides and their LRR-RLK receptors were already present in mosses [19] and that others have evolved with flowering plants [20]. Thus it is likely that residues of the RLK ectodomain important for peptide ligand binding have been conserved during evolution. Furthermore, peptide ligand binding is often facilitated by interaction with a co-receptor, commonly smaller membrane-bound LRR-RLKs with five LRRs [21].

Some LRR-RLKs are involved in pathogen defense and are activated by peptides derived from the pathogen, i.e., the FLAGELLIN SENSITIVE2 (FLS2) receptor binds flagellin 22 (flg22) of bacteria. To study this interaction, the lab of Professor Andrew Bent, University of Wisconsin, USA, developed a bioinformatics program to identify conserved residues, disregarding the six evenly spaced leucines (Leu, L) normally present in each repeat, structuring the LRR ectodomain [22]. Using this tool, we have recently identified conserved amino acids in the orthologues of the HAESA (HAE) and HAE-LIKE (HSL) receptors which in *A. thaliana* interact with the peptide IDA and regulate floral organ abscission and facilitate lateral root emergence [23]. These amino acids closely match the residues of HAE recently shown by structure determination to be directly interacting with the IDA peptide [24]. This suggests that the conservation mapping can be used to predict the peptide-interacting aa of the receptor. The method is dependent on correct identification of orthologues, as non-orthologous receptors are likely to differ in their ligand-binding amino acids. Therefore a phylogenetic analysis should precede conservation mapping.

Here we present a simplified method for the identification of orthologues of a receptor of interest and demonstrate the steps in the conservation mapping using HAE as an example. To indicate the reliability of the method, we have also analyzed the PSKR1 for which the structure of the LRR interacting with PSK and SERK co-receptor has been determined [25]. Also in this case, the mapping identified ligand-interacting residues. Thus we suggest that together with sequence alignment of ectodomains of orthologous receptors, LRR conservation mapping can identify residues interacting with peptide ligands. Comparison of conserved amino acids and structure information for HAE and PSKR1 indicates that the co-receptor-interacting residues also can be identified with this approach.

2 Materials

For this protocol some computer science skills are required: computer, terminal (command line), the Internet, and access to a server. This protocol is adapted from recently published work [20].

3 Methods

3.1 BLAST

1. BLASTP: Use the BLASTP [26] (protein BLAST) algorithm http://blast.ncbi.nlm.nih.gov/Blast.cgi with your protein of interest as a query. Change/use the following parameters on the BLASTP page:
   - Database: nonredundant protein sequences.
   - Organism: angiosperms (taxid:3398), if this is your target of interest.
   - Exclude: uncultured.
   - Max target sequences: 5000 (more than needed).
   - Expect threshold: 1e-180 (this will only return hits with an e-value of 0.0).
   - Algorithm: BLASTP.

2. On the BLAST Results page, select "ALL" in the "Descriptions" result and download either fasta or GenBank/pet (complete sequence).
3. Once downloaded you can edit the fasta header to only include the accession, source (species), and the gene/protein name; this can be performed manually or by scripting. Likewise, you can extract the required information from the GenBank file before converting it to a fasta file, again using scripts, e.g., http://app3.titan.uio.no/biotools/genbank2fasta.php or http://rocaplab.ocean.washington.edu/tools/genbank_to_fasta (some sequence analysis programs, e.g., Geneious, also offer this possibility). Save the fasta file to text (*see* **Note 1**) using, for example, TextWrangler (mac) or Notepad++ (win). This step will reduce the fasta header considerably, making subsequent steps easier (*see* **Note 1**).

3.2 Aligning Protein Sequences and Editing Alignment

1. Open MAFFT [27]multiple sequence alignment software, online: http://mafft.cbrc.jp/alignment/server/index.html. For MAFFT to work correctly, as outlined in this protocol, it needs to be opened in a browser with a java applet (e.g., Firefox). Paste or upload the protein sequences in the fasta text file into the input box and "submit" with default settings.
2. On the Results page, press "Refine dataset," scroll to the bottom of the page, and check if all sequences are loaded (essential if you have >1000 sequences).
3. Cluster the proteins, to remove redundant sequences, with CD-HIT [28]. Run CD-HIT with a 97% identity threshold, by pressing "ok" on the subsequent page (*see* **Note 2**).
4. Now "Clip the selected sequences and rebuild a tree" before pressing "Apply." "Submit" with default settings on the next page.
5. Now view the tree opened in the java applet ArchaeopteryxA (follow the instructions provided by MAFFT if ArchaeopteryxA didn't load). Here, you can select the node/clade you want to infer so only the sequences of interest will be used in downstream processing (*see* **Note 3**). For example, your original query sequence will be at the top, and if you kept the protein/gene name in the fasta header, it is easy to manually identify the ancestral node for a desired set of daughters. Select the ancestral node giving the desired daughter sequences and press "Realign the selected sequences" followed by "Apply" (*see* **Note 4**).
6. On the alignment page, align using default parameters, though if you have >200 sequences, the E-INS-I strategy is recommended to obtain a superior alignment. "Submit" and on the Results page download alignment in "fasta format" (*see* **Note 5**).

3.3 Phylogenetic Inference

The purpose with the phylogenetic analysis is to identify true orthologues of your receptor, i.e., receptors of other species that bind a peptide ligand homologous to a peptide activating your receptor of interest.

1. Open the alignment using MEGA (freely available at http://www.megasoftware.net/) [29] by Align → Edit/Build Alignment → Retrieve sequences from a file.
2. To infer a phylogeny in MEGA select: Phylogeny → Construct/Test Neighbor-Joining Tree… → and "open" the cured alignment in fasta format before selecting "Protein Sequences" and pressing "OK." On the subsequent page, change the following parameters before pressing "Compute":
   - Test of phylogeny: bootstrap method.
   - *No. of bootstrap replications*: 100.
   - Substitution model/method: JTT.
   - Rates among sites: gamma distributed (G).
   - *Gamma parameter*: 4.
   - Gaps/missing data treatment: pairwise deletion.
3. Use the resulting phylogeny to further edit the alignment to identify and filter duplicates using the following rule (Fig. 1): If a species forms a monophyletic or paraphyletic clade, duplicates

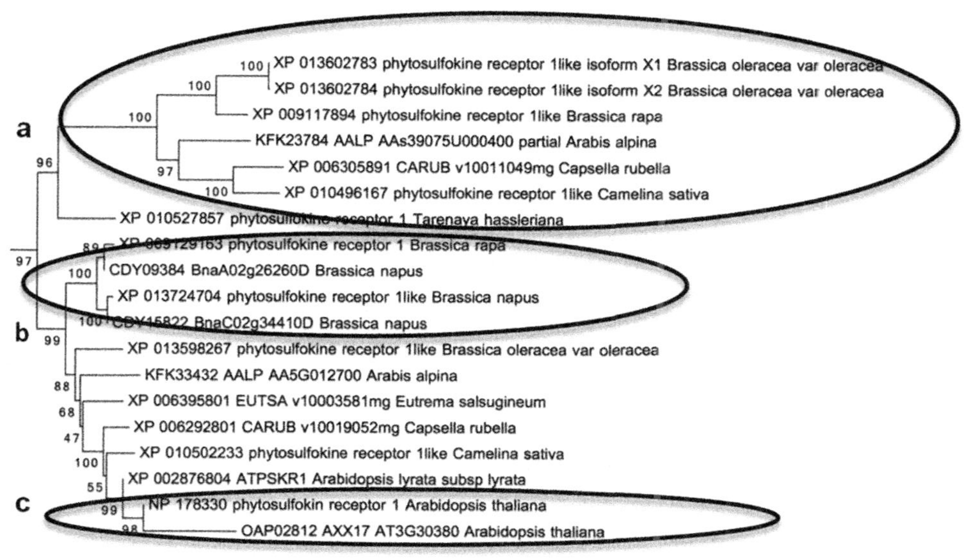

Fig. 1 Removal of duplicates and nonhomologous sequences exemplified with PSKR1. The tree depicts partial PSKR1 NJ phylogeny from Subheading 3.3, **step 2** in the protocol. The following rules are recommended: (**a**) of duplicates with equal branch lengths, retain those with the longest sequence; (**b**) remove clade depicting a possible genome duplication; and (**c**) remove species duplicate with a longer branch length

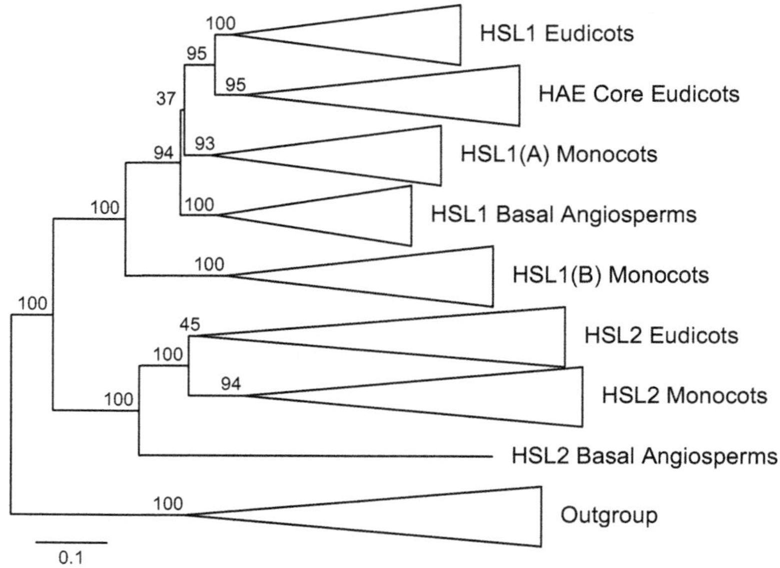

Fig. 2 Final NJ phylogeny of HSL LRR-RLK evolution within angiosperms after completion of Subheading 3.3. The tree has been collapsed at different taxonomical levels for ease of viewing. The final phylogeny can be used to identify and extract the sequences of interest from the final alignment that will be used for the generation of heatmaps (e.g., HAE in eudicots)

with a longer branch can be removed from the alignment. If branch lengths are equal, the taxon with the shortest sequence can be deleted from the alignment. Orders can often form multiple clades (often within a single grouping), representing genome duplications; however, the correct gene homolog forms a clade with sequences from additional orders and/or with a larger number of species representatives from within the order (*see* **Note 6**).

4. Repeat **step 2** to infer a final phylogeny with a single protein per species (*see* **Note 7**).
5. The final phylogeny can now be used to identify and extract the sequences of interest from the final alignment, e.g., HAE in eudicots or HSL2 in monocots. To do this simply delete those taxa that are not needed for further analysis before removing gaps from the alignment (w/o gaps) and then exporting the data as fasta format (Fig. 2) (*see* **Note 8**).

3.4 Generation of Heatmaps

1. Go to the Bent Lab LRR Conservation Mapping Program (http://144.92.198.58/main/main.php), find "Start RCM," press "GO," and thereafter "CLEAR" to remove the FLS2 sequences in the submission window.
2. Copy your sequences into the submission window in fasta format and summit; be sure that all unnecessary spaces and line shifts have been removed (*see* **Note 9**).

3. Aligned sequences will appear in the window. Tick off your species of interest, e.g., *Arabidopsis thaliana*. Press "LRR scan" and proceed to "LRR segmentation."
4. Manual adjustment of the LRR segments is normally needed, example given (Fig. 3) for the HAE receptor [23]. The program attempts to align all the LRRs, but irregular sequences must be dealt with. Remove sequences at the N-terminal end that are not structured as LRRs (e.g., the hydrophobic stretch of residues directing the protein to the secretory pathway). Also remove the transmembrane and intracellular domains C-terminal to the LRRs (*see* **Note 10**).
5. Sequences that have not been assigned to a LRR must be appended to the previous LRR (Fig. 3) (*see* **Note 11**). Furthermore it happens that short sequences can be assigned to two consecutive LRRs, and a choice has to be made. As a general advice, try to distribute the amino acids in question so that the length of the LRRs is similar, while at the same time, seek an optimal alignment of conserved L residues, or other conserved hydrophobic residues, i.e., valine (V), isoleucine (I), phenylalanine (F), or tyrosine (Y), relative to the position of such residues in the neighboring LRRs (Fig. 3).
6. When done press "Align sequences with ClustalW" to get to the intermediate alignment stage. Here it is possible to fix problems with sequences that we do not recognize as LRRs (Fig. 4a). LRRs heavily involved in ligand binding may often deviate slightly from the standard LRR. To be recognized in the conservation map, such deviating amino acid stretches should be reorganized as a new LRR.
7. Press "Proceed to realign". In the next appearing window, you can choose the starting amino acid for the LRRs. Normally the column with the conserved proline (P) residue is chosen (Fig. 4b).
8. The reordered alignment to be used for the LRR conservation mapping appears (Fig. 5). Make sure that no LRRs have been lost (*see* **Note 11**). The LRR alignment is now ordered to identify the positions of the conserved residues proline (P), leucine (L), asparagine (N), glycine (G), and isoleucine (I) which most commonly is PxxLxxLxxLxxLxLxxNxLxGxI (where x is any residue). Your alignment is normally a slightly modified version of this (e.g., for HAE and PSKR1 in Fig. 5a, b), and the pattern is to be submitted.
9. The final alignment appears and, after a click on the "Image" button, the heatmap (Fig. 5a, b). When the Pro column was chosen as a starting amino acid for the LRRs, the inner side of the concave ectodomain is on the right half of the heatmap, and the outer side is on the left [22].

```
>DUMMY(Do not change this line)
M

2          LYCLILLLCLSSTYLPSLSLNQDATI            LRR1
28         LRQAKLGLSDPAQSLSSWSDNNDVTPCKWL        LRR2

GVSCDATSNVVSVD

72         LSSFMLVGPFPSILCHLPSLHS                LRR1
94         LSLYNNSINGSLSADDFDTCHN                LRR2
116        LISLDLSENLLVGSIPKSLPFNLPN             LRR3
141        LKFLEISGNNLSDTIPSSFGEFRK              LRR4
165        LESLNLAGNFLSGTIPASLGNVTT              LRR5
189        LKELKLAYNLFSPSQIPSQLGNLTE             LRR6
214        LQVLWLAGCNLVGPIPPSLSRLTS              LRR7
238        LVNLDLTFNQLTGSIPSWITQLKT              LRR8
262        VEQIELFNNSFSGELPESMGNMTT              LRR9
286        LKRFDASMNKLTGKIPDNLNLLN               LRR10
309        LESLNLFENMLEGPLPESITRSKT              LRR11
333        LSELKLFNNRLTGVLPSQLGANSP              LRR12
357        LQYVDLSYNRFSGEIPANVCGEGK              LRR13
381        LEYLILIDNSFSGEISNNLGKCKS              LRR14
405        LTRVRLSNNKLSGQIPHGFWGLPR              LRR15
429        LSLLELSDNSFTGSIPKTIIGAKN              LRR16
453        LSNLRISKNRFSGSIPNEIGSLNGIIEISG        LRR17

AENDFSGEIPES

495        LVKLKQLSRLDLSKNQLSGEIPRELRGWKN        LRR1
525        LNELNLANNHLSGEIPKEVGILPV              LRR2
549        LNYLDLSSNQFSGEIPLELQNLK LRR3
572        LNVLNLSYNHLSGKIPPLYANKIYAHDFIG        LRR4

3: aSHIFT -18, 583, 601!

584        GKIPPLYANKIYAHDFIGNPGLCVD             LRR1
609        LDGLCRKITRSKNIGYVWILLTIFL             LRR2
634        LAGLVFVVGIVMFIAKCRKLRA                LRR3
656        LKSSTLAASKWRSFHKLHFSEHEIADCLDE        LRR4

KNVIGFGSSGKVYKVELRGGEVVAVKKLNKSVKGGDDEYSSDSLNRDVFAAE
VETLGTIRHKSIVRLWCCCSSGDCKLLVYEYMPNGSLADVLHGDRKGGVVLG
WPERLRIALDAAEGLSYLHHDCVPPIVHRDVKSSNILLDSDYGAKVADFGIA
KVGQMSGSKTPEAMSGIAGSCGYIAPEYVYTLRVNEKSDIYSFG

886        VVLLELVTGKQPTDSELGDKDMAKWVCTA         LRR1
915        LDKCGLEPVIDPKLDLKFKEEISKVIHIGL        LRR2
945        LCTSPLPLNRPSMRKVVIMLQEVSGAVPCS        LRR3

SPNTSKRSKTGGKLSPYYTEDLNSV
```

Fig. 3 LRR conservation mapping—the initial stage. Fasta format of orthologues of the receptor of interest (here eudicot orthologues of HAE) is submitted to give an initial LRR alignment that has to be edited manually. The non-LRR sequences must be removed, i.e., indicated in gray the N-terminal signal peptide (directing the

a

```
72   -----------LSSFMLVGP-FPSILCHLPSLHS---------------
94   ---------LSLYNNSINGSLSADDFDTCHN------------------
116  ------LISLDLSENLLVGS-IPKSLPFNLPN-----------------
141  ------LKFLEISGNNLSDT-IPSSFGEFRK------------------
165  ------LESLNLAGNFLSGT-IPASLGNVTT------------------
189  ------LKELKLAYNLFSPSQIPSQLGNLTE------------------
214  ------LQVLWLAGCNLVGP-IPPSLSRLTS------------------
238  ------LVNLDLTFNQLTGS-IPSWITQLKT------------------
262  ------VEQIELFNNSFSGE-LPESMGNMTT------------------
286  ------LKRFDASMNKLTGK-IPDNLNLLN-------------------
309  ------LESLNLFENMLEGP-LPESITRSKT------------------
333  ------LSELKLFNNRLTGV-LPSQLGANSP------------------
357  ------LQYVDLSYNRFSGE-IPANVCGEGK------------------
381  ------LEYLILIDNSFSGE-ISNNLGKCKS------------------
405  ------LTRVRLSNNKLSGQ-IPHGFWGLPR------------------
429  ------LSLLELSDNSFTGS-IPKTIIGAKN------------------
453  ------LSNLRISKNRFSGS-IPNEI-GSLNGIIEISGAENDFSGEIPES
495  LVKLKQLSRLDLSKNQLSGE-IPRELRGWKN------------------
525  ------LNELNLANNHLSGE-IPKEVGILPV------------------
549  ------LNYLDLSSNQFSGE-IPLELQNLK-------------------
572  ------LNVLNLSYNHLSGK-IP--------------------------
```

b

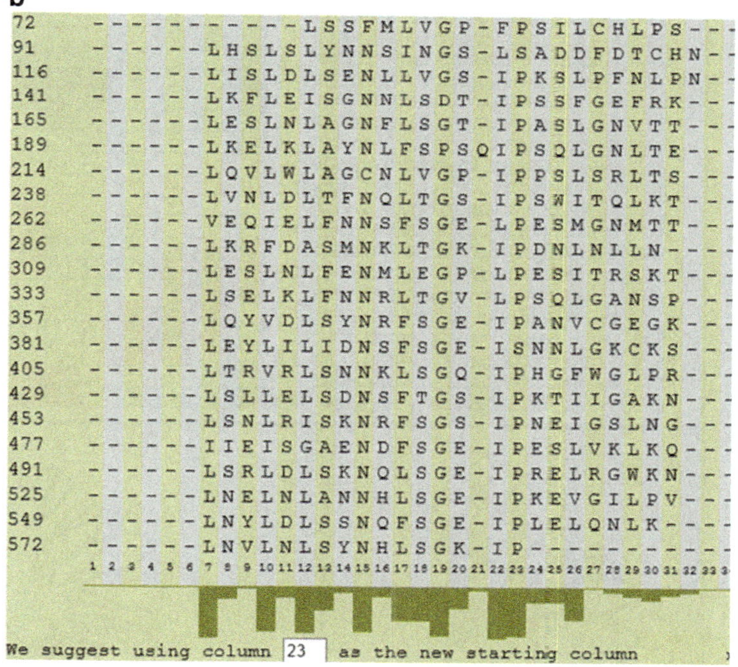

Fig. 4 LRR conservation mapping—the intermediate stage After manual adjustment and ClustalW multiple alignment of the LRRs, an improved alignment is generated (**a**), and thereafter (**b**) a new line break, usually before the highly conserved proline (P), can be selected

Fig. 3 (continued) receptor to the endoplasmatic reticulum for export to the plasma membrane) and C-terminally to the LRRs the transmembrane and kinase domains. Nonaligned amino acid stretches (*boxed*) must be moved to the end of the previous LRR (*arrow*), and sequences indicated to fit in two neighboring LRRs (*underlined*) must be assigned to one or the other LRR

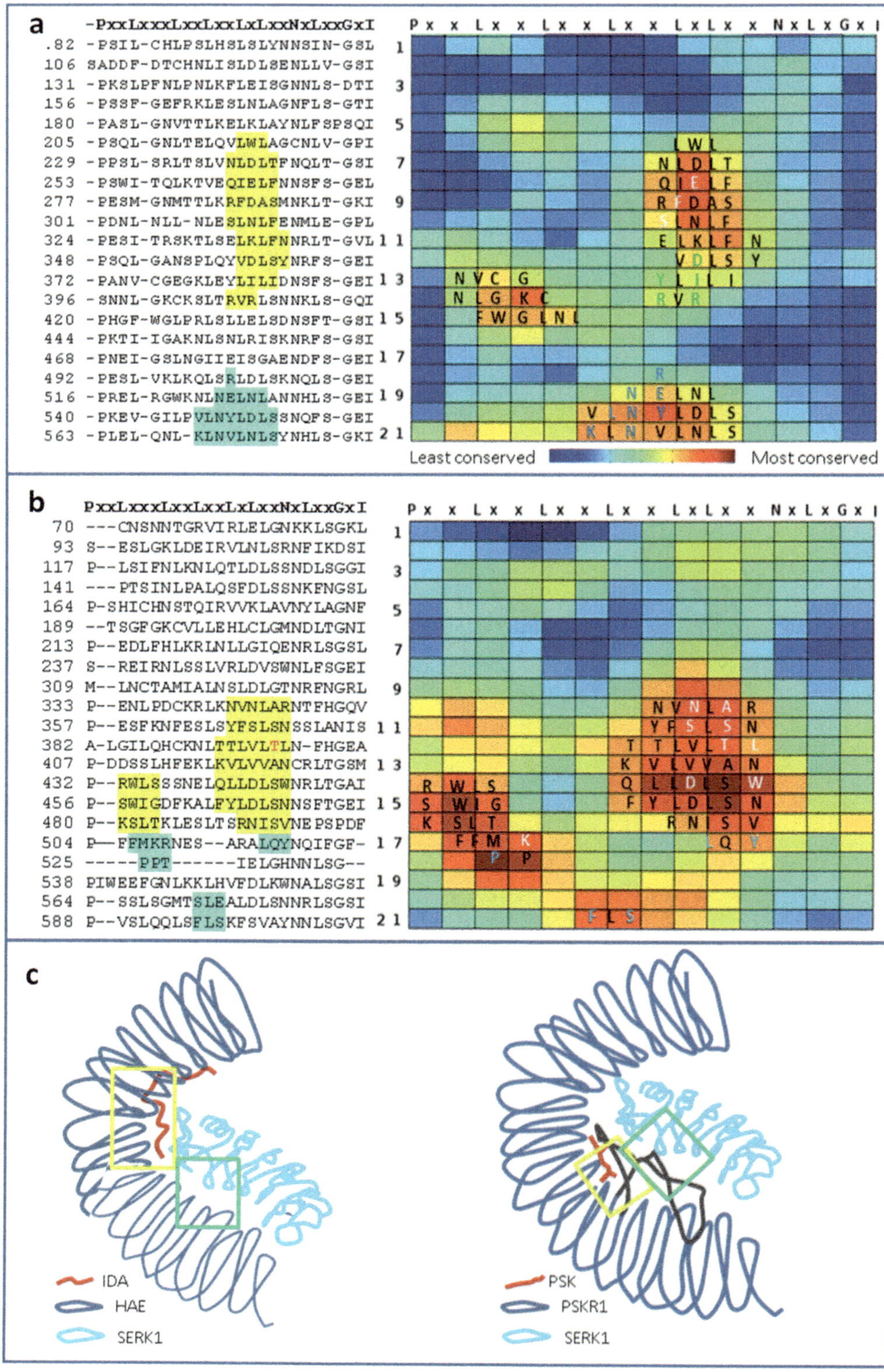

Fig. 5 LRR conservation mapping—heatmap. (**a**) Final LRR alignment and heatmap for the HAE receptor. (**b**) LRR alignment and heatmap for PSKR1 of eudicots. (**c**) Cartoons of the HAE and PSKR1 with ligands and co-receptors based on the published structures [24, 25]. *Yellow-* and *blue*-marked amino acids in the alignments

10. Identify the conserved amino acids by comparing the aligned sequence and the heatmap (e.g., for HAE Fig. 5a, c and for PSKR1, Fig. 5b, c).

4 Notes

1. Additional sequence data from other databases (sources) can be added to the fasta text file in this step (e.g., Phytozome, Ancestral Angiosperm Genome Project (AAGP), Comparative Genomics (CoGe)).
2. In the protein databases, there are often multiple accessions of the same protein, i.e., the protein encoded by a single gene. Such duplicates must be removed. The threshold is dependent on both the aim/goal and the level of sequence conservation; however, we have found 97% to cluster most proteins at the species level.
3. We recommend that an additional clade, representing an out-group, is kept for phylogenetic inference. The inclusion of an out-group will allow for easier identification and removal of non-homologous sequences in subsequent steps.
4. This step can be refined by using the "Search (A) and (B)" options in ArchaeopteryxA, identifying species with a broad taxonomical range, e.g., *Arabidopsis* and *Oryza*. Note, however, that A and B need to be "reset" before selecting the correct ancestral node; otherwise, any As and Bs external to this selection will also be included in downstream processing.
5. Subheading 3.2, **steps 1–6** reduce the dataset considerably; however, it is not able to remove all gene paralogs/duplicates for each species. Therefore we need to proceed with the next steps to infer a topology to eradicate any additional deviant protein copies with the intension to retain only true orthologues.
6. We recommend running the phylogeny step (Subheading 3.3, **step 2**) again after a first round of sequence removal to make it easier to identify additional duplicates.
7. It is important to eradicate multiple gene copies for each species to avoid sequence overrepresentation in the subsequent steps.

Fig. 5 (continued) (**a**) and (**b**), and *boxes* in the structure drawings in (**c**), indicate conserved regions interacting with the ligands and co-receptors, respectively. Amino acids in the heatmaps in *white* and *green* have been shown to interact directly with amino acids of the ligand, while those in *blue* interact directly with the co-receptor [24, 25]

8. It is important that the alignment is in a sequential fasta format for the LRR conservation mapping steps. The LRR Conservation Mapping Program will not work if there are additional spaces before, between or after the different sequences, or if there are line shifts at the end of the lines.
9. It is advisable not to have more than 50 sequences, as the program work slowly when there are many sequences to be aligned.
10. Protein accessions or the corresponding gene accession often provide information on the position of the domains along the protein sequence.
11. If not reorganized such sequences tend to be lost from the final conservation map.

References

1. Matsubayashi Y (2014) Posttranslationally modified small-peptide signals in plants. Annu Rev Plant Biol 65:385–413
2. Bartels S, Boller T (2015) Quo vadis, pep? Plant elicitor peptides at the crossroads of immunity, stress, and development. J Exp Bot 66(17):5183–5193
3. Motomitsu A, Sawa S, Ishida T (2015) Plant peptide hormone signalling. Essays Biochem 58:115–131
4. Lease KA, Walker JC (2006) The Arabidopsis unannotated secreted peptide database, a resource for plant peptidomics. Plant Physiol 142(3):831–838
5. Butenko MA, Vie AK, Brembu T, Aalen RB, Bones AM (2009) Plant peptides in signalling: looking for new partners. Trends Plant Sci 14(5):255–263
6. Wheeler JI, Irving HR (2012) Plant peptide signaling: an evolutionary adaptation. In: Gehring CA, Irving HR (eds) Plant signaling peptides, Signaling and communication in plants, vol 16. Springer-Verlag, Berlin, Heidelberg. doi:10.1007/978-3-642-27603-3_1
7. Yu F, Luan S (2016) Peptide signaling in plants: finding partners is the key. Cell Res 26:755–756. doi:10.1038/cr.2016.73
8. Shiu SH, Bleecker AB (2001) Receptor-like kinases from Arabidopsis form a monophyletic gene family related to animal receptor kinases. Proc Natl Acad Sci U S A 98(19):10763–10768
9. Etchells JP, Turner SR (2010) The PXY-CLE41 receptor ligand pair defines a multi-functional pathway that controls the rate and orientation of vascular cell division. Development 137(5):767–774
10. Kondo T, Yokomine K, Nakagawa A, Sakagami Y (2010) Analogs of the CLV3 peptide: synthesis and structure–activity relationships focused on proline residues. Plant Cell Physiol 52(1):30–36. doi:10.1093/pcp/pcq146
11. Meng L, Feldman L (2010) CLE14/CLE20 peptides may interact with CLAVATA2/CORYNE receptor-like kinases to irreversibly inhibit cell division in the root meristem of Arabidopsis. Planta 232(5):1061–1074
12. Wang G, Fiers M (2010) CLE peptide signaling during plant development. Protoplasma 240(1):33–43
13. Stahl Y, Simon R (2012) Peptides and receptors controlling root development. Philos TransRoyal Soc B-Biol Sci 367(1595):1453–1460
14. Qiang Y, Wu JB, Han HB, Wang GD (2013) CLE peptides in vascular development. J Integr Plant Biol 55(4):389–394
15. Richardson LG, Torii KU (2013) Take a deep breath: peptide signalling in stomatal patterning and differentiation. J Exp Bot 64(17):5243–5251
16. Ogilvie HA, Imin N, Djordjevic MA (2014) Diversification of the C-TERMINALLY ENCODED PEPTIDE (CEP) gene family in angiosperms, and evolution of plant-family specific CEP genes. BMC Genomics 15(1):1–15
17. Ingram G, Gutierrez-Marcos J (2015) Peptide signalling during angiosperm seed development. J Exp Bot 66(17):5151–5159
18. Yamaguchi YL, Ishida T, Sawa S (2016) CLE peptides and their signaling pathways in plant development. J Exp Bot 67(16):4813–4826

19. Miwa H, Tamaki T, Fukuda H, Sawa S (2009) Evolution of CLE signaling: origins of the CLV1 and SOL2/CRN receptor diversity. Plant Signal Behav 4(6):477–481

20. Stø IM, Orr RJS, Fooyontphanich K, Jin X, Knutsen JMB, Fischer U, Tranbarger TJ, Nordal I, Aalen RB (2015) Conservation of the abscission signaling peptide IDA during angiosperm evolution: withstanding genome duplications and gain and loss of the receptors HAE/HSL2. Front Plant Sci 6:931. doi:10.3389/fpls.2015.00931

21. aan den Toorn M, Albrecht C, de Vries S (2015) On the origin of SERKs: bioinformatics analysis of the somatic embryogenesis receptor kinases. Mol Plant 8(5):762–782

22. Helft L, Reddy V, Chen X, Koller T, Federici L, Fernandez-Recio J, Gupta R, Bent A (2011) LRR conservation mapping to predict functional sites within protein leucine-rich repeat domains. PLoS One 6(7):e21614

23. Aalen RB, Wildhagen M, Stø IM, Butenko MA (2013) IDA: a peptide ligand regulating cell separation processes in Arabidopsis. J Exp Bot 64(17):5253–5261

24. Santiago J, Brandt B, Wildhagen M, Hohmann U, Hothorn LA, Butenko MA, Hothorn M (2016) Mechanistic insight into a peptide hormone signaling complex mediating floral organ abscission. eLife 5:e15075. doi:10.7554/eLife.15075

25. Wang J, Li H, Han Z, Zhang H, Wang T, Lin G, Chang J, Yang W, Chai J (2015) Allosteric receptor activation by the plant peptide hormone phytosulfokine. Nature 525(7568):265–268

26. Altschul SF, Gish W, Miller W, Myers EW, Lipman DJ (1990) Basic local alignment search tool. J Mol Biol 215(3):403–410

27. Katoh K, Standley DM (2013) MAFFT multiple sequence alignment software version 7: improvements in performance and usability. Mol Biol Evol 30(4):772–780

28. Li W, Jaroszewski L, Godzik A (2001) Clustering of highly homologous sequences to reduce the size of large protein databases. Bioinformatics 17(3):282–283

29. Kumar S, Stecher G, Tamura K (2016) MEGA7: molecular evolutionary genetics analysis version 7.0 for bigger datasets. Mol Biol Evol 33(7):1870–1874

Part IV

Identification of Enzymatic Functions and Interacting Partners of PRKs

Chapter 10

Co-immunoprecipitation of Plant Receptor Kinases

María Laura Barberini and Jorge P. Muschietti

Abstract

In order to comprehend the function of a particular protein, identification of the interacting protein partners is a useful approach. Co-immunoprecipitation (Co-IP) is employed to test physical interactions between proteins. Specific antibodies or antibodies against tagged versions can be used to immunoprecipitate the proteins. In this chapter, we describe a method to carry out Co-IP using recombinant membrane proteins expressed in yeast microsomal fractions.

Key words Yeast, Co-immunoprecipitation, Protein–protein interaction, Antibody, Protein interaction

1 Introduction

Interpreting how cell receptors perceive signals and transduce them into the cytoplasm is a necessary step to understand how cells interact and respond to a fluctuating environment. Co-immunoprecipitation (Co-IP) has become a valuable approach for analyzing protein–protein interactions and posttranslational modifications or to isolate protein complexes. This method is based on the ability of a protein to interact with other proteins. The protein complex is incubated with an antibody against the first protein and pulled down by protein A Sepharose beads. The precipitated proteins are separated by SDS-PAGE, immunoblotted, and developed using an antibody against any of the other proteins. The presence of a band concludes interaction between the corresponding proteins.

Co-IP has been extensively used in plants to analyze interaction between receptor-like kinases (RLKs) that are receptor proteins with transmembrane domains, variable N-terminal extracellular domains and C-terminal intracellular kinases [1–7].

Here we present in detail the co-immunoprecipitation protocol we successfully used to analyze interactions between two pollen-specific receptor-like kinases from *Solanum lycopersicum*

Reidunn Birgitta Aalen (ed.), *Plant Receptor Kinases: Methods and Protocols*, Methods in Molecular Biology, vol. 1621, DOI 10.1007/978-1-4939-7063-6_10, © Springer Science+Business Media LLC 2017

(LePRK1 and LePRK2) expressed in yeast microsomal fractions [6–8]. This protocol can be modified to analyze interactions in planta or when transiently expressed in other biological systems.

2 Materials

2.1 Yeast Material

Yeast microsomal membranes that express the two proteins whose interaction is analyzed. In our case, the two pollen-specific receptor-like kinases from *Solanum lycopersicum* (LePRK1 and LePRK2) [1] (*see* Chapter 2).

2.2 Reagents

Non-denaturing immunoprecipitation (ND-IP) buffer: 50 mM Tris–HCl, pH 8.0, 150 mM NaCl, and 0.5% (v/v) Nonidet P-40.

Protease inhibitor set: cOmplete ULTRA Tablets, EDTA-free (Fisher).

Protein A Sepharose: Stock 3% in ND-IP buffer (protein A Sepharose CL-4B beads; Sigma).

Laemmli SDS-PAGE buffer 1×: 50 mM Tris–HCl, pH 8.0, 2% (v/v) SDS, 10% (v/v) glycerol, 5% (v/v) β-mercaptoethanol, 0.01% (w/v) bromophenol blue.

Tris-buffered saline (TBS) 1×: 50 mM Tris–HCl, pH 7.5, and 150 mM NaCl.

Tris-buffered saline with Triton X-100 (TBST) buffer: 1× TBS with 0.2% (v/v) Triton X-100.

Blocking buffer: 6% (w/v) nonfat dry milk and 4% (w/v) glycine in TBST.

Antibody dilution buffer: 0.3% (w/v) nonfat dry milk and 0.3% (w/v) glycine in TBST.

3 Methods

3.1 Co-immuno-precipitations

1. Resuspend the yeast microsomal membranes (750 μg of protein) in non-denaturing immunoprecipitation (ND-IP) buffer by stirring on a magnetic stirrer at 4 °C for 1 h, and adjust to a final protein concentration of 15 μg/μl.
2. Incubate the resuspended yeast microsomal membranes with 2 μl of the first antibody against the first protein of the interactive pair (in our case LePRK2 antibody) (*see* **Note 1**). Rotate sample tubes for 2 h at 4 °C.
3. Centrifuge the mixture at 10,000 × *g* for 10 min at 4 °C. Keep the supernatant.
4. Add 100 μl of 10% (v/v) protein A Sepharose (pre-equilibrated in ND-IP buffer) to the supernatant, and incubate for 1 h at 4 °C (*see* **Note 2**).

5. Pellet the beads at a low-speed setting for 10 s in the centrifuge. Remove supernatant.
6. Wash the beads twice with ND-IP buffer and spin as **step 5**.

3.2 SDS-PAGE and Immunoblotting

1. Resuspend the beads in 60 μl of 1× Laemmli SDS-PAGE buffer and boil for 3 min.
2. After pelleting the beads, retain the supernatant.
3. Separate the proteins by SDS-PAGE, blot to nitrocellulose, and immunoblot (*see* **Note 3**).
4. Incubate the membranes in blocking buffer for 30 min.
5. Develop the immunoblot with an antibody against the second protein, the candidate interacting protein (in our case LePRK1 antibody) (*see* **Note 1**) diluted to 1:1000 in antibody dilution buffer for 1 h with shaking at room temperature.
6. After six washes of 10 min each with TBST, the membranes are incubated for 1 h with the corresponding polyclonal secondary antibodies conjugated with horseradish peroxidase (Amersham Biosciences), washed, and developed by using the enhanced chemiluminescence kit (Amersham Biosciences).
7. The presence of a band corresponding to the second protein (in our case LePRK1) after immunoprecipitation with an antibody against the first one (e.g., LePRK2) infers that they associate in the plasma membrane (*see* **Notes 4–6**).

4 Notes

1. We used our own antibodies against LePRK1 and LePRK2 [6–8]. Alternatively, commercial antibodies against epitope tags could be used.
2. If protein A has low affinity for your specific antibody, it is recommended to add a secondary antibody to increase the protein recovery.
3. Include in the gel 40–60 μg of the yeast microsomal fractions as the "Input" sample.
4. A negative control should be included in each set of immunoprecipitations. Use a strain that only expresses one of the interacting proteins (in our case LePRK1), and for the immunoprecipitation use the antibody against the other candidate interacting protein (in our case LePRK2). Develop the blot using an antibody against the first protein (in our case LePRK1). The presence of a band implies nonspecific background binding to the beads or cross-reaction between the antibodies.

5. It is also recommended to carry out the procedure in the absence of the first antibody or using normal serum.
6. To diminish nonspecific interactions, modify the ionic strength of the IP buffer, and decrease the amount of primary antibody or preclear the lysates.

Acknowledgments

This work was supported by grants to J.M. (UBACyT, PICT2012, PICT2014, and PICT2015).

References

1. Xu T, Dai N, Chen J, Nagawa S, Cao M, Li H, Zhou Z, Chen X, De Rycke R, Rakusová H, Wang W, Jones AM, Friml J, Patterson SE, Bleecker AB, Yang Z (2014) Cell surface ABP1-TMK auxin-sensing complex activates ROP GTPase signaling. Science 343(6174): 1025–1028
2. Meyer MR, Shah S, Zhang J, Rohrs H, Rao AG (2015) Evidence for intermolecular interactions between the intracellular domains of the Arabidopsis receptor-like kinase ACR4, its homologs and the Wox5 transcription factor. PLoS One 10(3):e0118861. doi:10.1371/journal.pone.0118861
3. Lu D, Wu S, Gao X, Zhang Y, Shan L, He P (2010) A receptor-like cytoplasmic kinase, BIK1, associates with a flagellin receptor complex to initiate plant innate immunity. Proc Natl Acad Sci U S A 107(1):496–501
4. Li J, Wen J, Lease KA, Doke JT, Tax FE, Walker JC (2002) BAK1, an Arabidopsis LRR receptor-like protein kinase, interacts with BRI1 and modulates brassinosteroid signaling. Cell 110(2):213–222
5. Nam KH, Li J (2002) BRI1/BAK1, a receptor kinase pair mediating brassinosteroid signaling. Cell 110(2):203–212
6. Wengier D, Valsecchi I, Cabanas ML, Tang WH, McCormick S, Muschietti J (2003) The receptor kinases LePRK1 and LePRK2 associate in pollen and when expressed in yeast, but dissociate in the presence of style extract. Proc Natl Acad Sci U S A 100(11):6860–6865
7. Salem TM, Barberini ML, Wengier DL, Cabanas ML, de Paz P, Muschietti J (2012) Oligomerization studies show that the kinase domain of the tomato pollen receptor kinase LePRK2 is necessary for interaction with LePRK1. Plant Physiol Biochem 53:40–45
8. Muschietti J, Eyal Y, McCormick S (1998) Pollen tube localization implies a role in pollen-pistil interactions for the tomato receptor-like protein kinases LePRK1 and LePRK2. Plant Cell 10(3):319–330

Chapter 11

Autophosphorylation Assays Using Plant Receptor Kinases Synthesized in Cell-Free Systems

Keiichirou Nemoto, Akira Nozawa, Satoshi Yamanaka, Shunsuke Nomura, Kohki Kido, and Tatsuya Sawasaki

Abstract

The wheat germ cell-free protein synthesis system has a significant advantage for high-throughput production of eukaryotic multidomain proteins in a folded state. In this chapter, we describe two kinds of methods for performing autophosphorylation assay of plant receptor kinases (PRKs) by using the wheat cell-free system. One is an in vitro kinase assay performed using biotin-streptavidin affinity purification technology, and the other is a luminescence-based high-throughput assay for autophosphorylation analysis. We anticipate that our cell-free-based methods might facilitate the characterization of plant PRKs.

Key words Wheat germ cell-free protein synthesis system, AlphaScreen, Autophosphorylation, Plant receptor kinase

1 Introduction

A typical plant receptor kinase (PRK) consists of an extracellular domain, a transmembrane domain, and a cytoplasmic protein kinase domain. PRKs recognize a wide range of signals or stimuli by the extracellular domain and activate the downstream signaling pathway by phosphorylating target proteins. Many studies on mammalian receptor kinases have revealed that autophosphorylation of the cytoplasmic domain is an important aspect of regulatory systems for conformational change, homo- or hetero-oligomerization, and activation [1]. In plants, the protein kinase domains of several PRKs such as BRASSINOSTEROID-INSENSITIVE 1 (BRI1) and CHITIN ELICITOR RECEPTOR KINASE 1 (CERK1) were found to have ligand-induced homo-/hetero-oligomerization and autophosphorylation activity [2, 3]. Thus, the autophosphorylation activity of PRKs is one of the important aspects to elucidate the activation mechanism and downstream signaling pathways. However, the biochemical

Reidunn Birgitta Aalen (ed.), *Plant Receptor Kinases: Methods and Protocols*, Methods in Molecular Biology, vol. 1621, DOI 10.1007/978-1-4939-7063-6_11, © Springer Science+Business Media LLC 2017

characterization of PRKs has been hindered by the difficulties in obtaining sufficient quantities of functionally active recombinant protein.

In general, preparing sufficient amount of functional PRKs is difficult because of its low abundance in natural sources and the relatively low successful rate of obtaining them in recombinant expression systems owing to the inhibitory effect on host cell physiology. To overcome these difficulties, we developed wheat germ cell-free protein synthesis system to produce PRKs (*see* Chapter 4). In many cases, the wheat cell-free system can synthesize full-length PRKs as soluble proteins without requiring additional supplements such as surfactants. In addition, we developed a biotinylation tag system of the target proteins and could rapidly and efficiently purify target proteins by using the biotin-streptavidin affinity purification technology [4, 5]. This method allowed the rapid purification of recombinant PRKs and was useful for conducting autophosphorylation analysis [6].

Ideally, highly purified proteins should be used for biochemical analysis. However, the purification of recombinant proteins is often difficult and time-consuming. PRKs have hydrophobic transmembrane domains and are especially prone to aggregation in solution. In order to address these issues, we have already established highly sensitive and specific methods for autophosphorylation analysis of protein kinases without requiring purification steps by combining the wheat cell-free system and a luminescence system "AlphaScreen" by using the anti-phosphoamino acid antibody [6, 7]. AlphaScreen, amplified luminescent proximity homogeneous assay, is a bead-based proximity-dependent chemical energy transfer luminescent assay and can be used to detect the interaction of analyte pairs [8]. We applied this approach and performed autophosphorylation activity profiling of 277 biotinylated protein kinases and detected the autophosphorylation activity of 53 PRKs [6, 7]. These results indicated that our cell-free system is a promising tool for the production and functional analysis of PRKs. In addition, the methodology combining wheat cell-free system and AlphaScreen technology is available for functional analysis or substrate screening of post-modification enzymes such as protein kinases, protein phosphatases, proteases, and ubiquitin ligases [9–12].

2 Materials

2.1 An In Vitro Kinase Assay for Biotinylated PRKs

1. Streptavidin MagneSphere paramagnetic particles (SA-PMPs; 1 mg/ml; Promega, Madison, WI).
2. Wash buffer A: 50 mM Tris–HCl (pH 7.4), 150 mM NaCl, 0.5% Tween 20.

3. Wash buffer B: 50 mM Tris–HCl (pH 7.4), 150 mM NaCl, 1% Tween 20, 0.1% sodium lauryl sulfate (SDS), 5 mM EDTA.
4. Magnetic stand.
5. 1× In vitro kinase buffer: 50 mM Tris–HCl (pH 7.5), 100 mM potassium acetate, 10 mM magnesium chloride, 1 mM dithiothreitol, 100 μM ATP. For autoradiography analysis, add 1–50 μCi of [gamma-^{32}P]-ATP in the 1× in vitro kinase buffer.
6. 2× SDS sample buffer: 125 mM Tris–HCl (pH 6.8), 4% SDS, 20% glycerol, 0.01% bromophenol blue, 10% β-mercaptoethanol. The β-mercaptoethanol must be added before use.

2.2 Autophosphorylation Analysis of Biotinylated PRKs by Using the Luminescence System "AlphaScreen"

1. AlphaScreen IgG (protein A) detection kit (PerkinElmer, Boston, MA). This kit contains the streptavidin-coated donor beads and protein A-coated acceptor beads. Store at 4 °C.
2. Anti-phospho-serine/threonine antibody (anti-pSer/pThr antibody; Upstate Biotechnology, Lake Placid, NY). Store at −30 °C.
3. Anti-phosphotyrosine antibody, clone 4G10 (anti-pTyr antibody; EMD Millipore, Billerica, MA). Store at −30 °C.
4. 10× BSA: 10 mg/ml bovine serum albumin (BSA) in sterile water. Store at −30 °C.
5. 5× In vitro kinase buffer: 250 mM Tris–HCl (pH 7.5), 500 mM potassium acetate, 50 mM magnesium chloride, 5 mM dithiothreitol, 500 μM ATP. Store at −80 °C.
6. 10× Detection buffer: 1 M Tris–HCl (pH 8.0), 0.1% Tween 20, 1% Brij 35, 50 mM EDTA (*see* **Note 1**). Store at −30 °C.
7. White 384-well plates for luminescence, fluorescence, and colorimetric assays.
8. EnVision Multilabel Plate Reader (PerkinElmer, Boston, MA).
9. Plate-sealing films.

3 Methods

3.1 The In Vitro Kinase Assay of Biotinylated PRKs Performed Using Biotin-Streptavidin Affinity Purification Technology

In this section, we describe the protocol for the purification of biotinylated PRKs and for the in vitro kinase assay. This method can be used for immunoblot analysis by using anti-phosphoamino acid antibody or autoradiography analysis by using [gamma-^{32}P]-ATP.

1. Resuspend SA-PMPs by gently inverting or pipetting the vial until the particles are completely dispersed.
2. Transfer 10 μl SA-PMPs to the tube (one-fifth volume of biotinylated PRK).

3. Place the tube on a magnetic stand to separate the SA-PMPs from the solution (approximately 30 s).
4. Wash SA-PMPs three times, each with 100 μl of wash buffer A.
5. Resuspend SA-PMPs in 10 μl wash buffer A (equal original volume of SA-PMPs).
6. Add 50 μl biotinylated PRK that was synthesized using the wheat germ cell-free protein synthesis system (*see* **Note 2**).
7. Remove the tube from the magnetic stand.
8. After the mixture is gently mixed by pipetting, incubate it at room temperature for 15 min.
9. Place the tube on the magnetic stand. Next, carefully remove the supernatant.
10. Wash SA-PMPs three times, each with 100 μl of wash buffer A (twice the original volume of biotinylated-RPK).
11. Remove the tube from the magnetic stand.
12. Add 50 μl of 1× in vitro kinase buffer and gently mix.
13. Incubate at 25 °C for approximately 1 h (for autoradiography analysis, incubate at 25 °C for 10–30 min).
14. Place the tube on the magnetic stand. Subsequently, carefully remove the supernatant.
15. Wash SA-PMPs four times, each with 100 μl of wash buffer B (twice the original volume of biotinylated PRK).
16. After the supernatant is completely removed, add 50 μl of 2× SDS sample buffer.
17. Incubate at 100 °C for 5 min.
18. Separate 25 μl of denatured proteins by SDS polyacrylamide gel electrophoresis (SDS-PAGE). Next, detect the phosphorylated proteins by using a typical immunoblot analysis with anti-phosphoamino acid antibody (Fig. 1) or autoradiography analysis.

3.2 Autophosphorylation Analysis of Biotinylated PRKs Performed Using Luminescence System "AlphaScreen"

In this section, we describe how to perform autophosphorylation analysis of biotinylated PRKs by using the luminescence system "AlphaScreen" and the anti-phosphoamino acid antibody (Fig. 2). The autophosphorylation of PRKs can be analyzed by synthesizing biotinylated PRKs by using the wheat cell-free system (*see* Chapter 4). Next, two types of AlphaScreen beads, streptavidin-coated donor beads and protein A-coated acceptor beads, and anti-phosphoamino acid antibody are added to the reaction mixture. Biotinylated PRK binds to streptavidin on the donor beads with an extremely specific and high affinity. The protein A-coated acceptor beads are combined with autophosphorylation biotinylated PRK through the anti-phosphoamino acid antibody. Finally, autophosphorylated

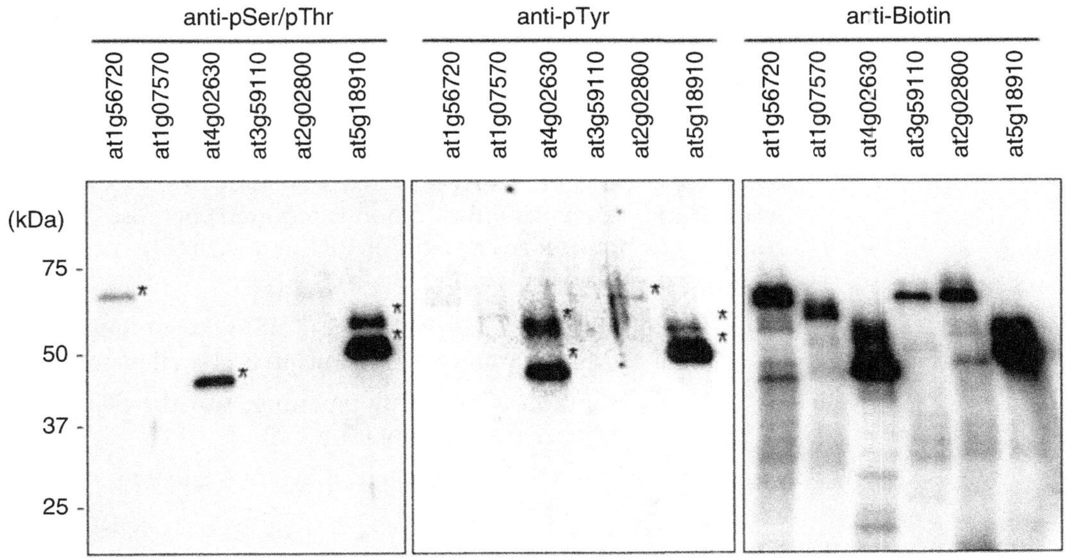

Fig. 1 Autophosphorylation analysis by using immunoblot with anti-phosphoamino acid antibody. Six biotinylated PRKs were purified using the biotin-streptavidin affinity purification technology. After the biotinylated PRKs were incubated in kinase buffer, phosphorylation was detected by immunoblot analysis with anti-pSer/pThr antibody (*left panel*) and anti-pTyr antibody (4G10; *middle panel*). Biotinylated PRKs were detected using anti-biotin antibody (*right panel*). Asterisk indicates phosphorylation signal

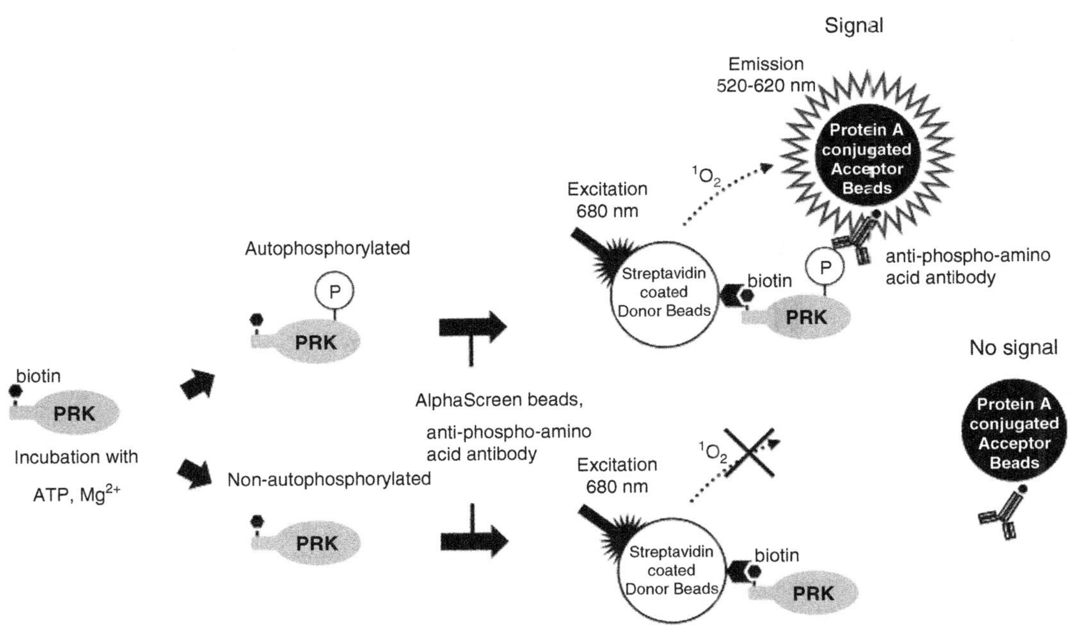

Fig. 2 Diagrammatic representation of the autophosphorylation analysis by using the luminescence system "AlphaScreen." When autophosphorylation of biotinylated PRK occurs, the complex, including two kinds of AlphaScreen beads, is formed, and autophosphorylation signal can be detected by luminescence (*upper right*). However, when autophosphorylation does not occur, the signal cannot be detected (*lower right*)

biotinylated PRK forms a big complex with two kinds of beads through the antibody and streptavidin. After illumination at 680 nm, the donor beads convert ambient oxygen to singlet oxygen ($^{1}O_2$). When PRKs have autophosphorylation activity, singlet oxygen is transferred across to activate the acceptor beads and subsequently emits light at 520–620 nm. This assay can be performed using crude protein (no purification is required) because biotinylation tag or phosphoamino acid of PRK is specifically recognized by the two kinds of beads (Fig. 3a, b).

1. Add 1 μl of biotinylated PRK, 1.5 μl of 5× kinase buffer, and 12.5 μl of Milli-Q water (total 15 μl) to 384-well plates.
2. After the solution is mixed by pipetting, seal the plates with sealing films to prevent evaporation.
3. After the plates are centrifuged, they are incubated at 25 °C for 1 h.

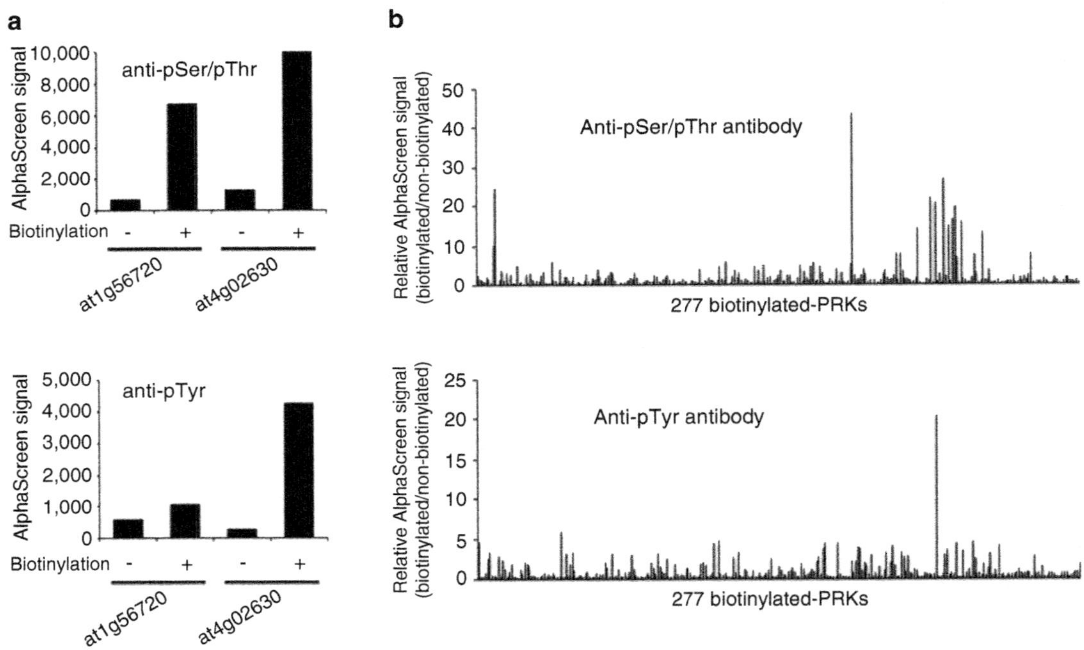

Fig. 3 Autophosphorylation analysis of biotinylated PRKs by using the luminescence system "AlphaScreen." (**a**) Autophosphorylation analysis of at1g56720 and at4g02630 performed using anti-pSer/pThr antibody or anti-pTyr antibody. Biotinylated or non-biotinylated protein was synthesized using the wheat cell-free system. Immunoblot analysis images of at1g56720 and at4g02630 are shown in Fig. 1. (**b**) Autophosphorylation analysis of 277 biotinylated PRK array performed using anti-pSer/pThr antibody (*upper panel*) or anti-pTyr antibody (*lower panel*). All data are the average of two independent experiments, and the background was controlled for each experiment by using the relevant non-biotinylated PRKs

4. Prepare the detection solution in the following order in a tube:

10× Detection buffer	2.5 μl
10× BSA (10 mg/ml)	2.5 μl
Anti-pSer/pThr antibody or anti-pTyr antibody	0.025 μl
Streptavidin-coated donor beads	0.1 μl
Protein A-coated acceptor beads	0.1 μl
Milli-Q water	4.775 μl

The total volume of detection solution is 10 μl per sample. Subsequent operations should be performed in a dark room under green light because AlphaScreen beads are photosensitive.

5. Add 10 μl of detection solution (final volume at 25 μl).
6. Seal the plates with sealing films and wrap them in aluminum foil. Next, centrifuge the plates.
7. Incubate at 25 °C for 1 h.
8. After the seal is peeled off, measure the luminescence signal by using an EnVision Multilabel Plate Reader.

4 Notes

1. AlphaScreen analysis involves the use of non-denaturation recombinant protein. The three-dimensional structure of the protein often influences the accessibility of the antibody. It is usually improved by adding Brij 35 (approximately 1%) or NaCl (approximately 150 mM) to the final reaction solution.
2. The synthetic amount might differ depending on the translation methods and protein, but the wheat germ cell-free protein synthesis system can produce nanogram to milligram amounts of protein per reaction. The amount of synthesized protein to bind to the beads can be changed. The amount of protein required for the experiment can be determined experimentally.

References

1. Lemmon MA, Schlessinger J (2010) Cell signaling by receptor tyrosine kinases. Cell 141:1117–1134
2. Faulkner C (2013) Receptor-mediated signaling at plasmodesmata. Front Plant Sci 4:521
3. Lozano-Durán R, Zipfel C (2015) Trade-off between growth and immunity: role of brassinosteroids. Trends Plant Sci 20:12–19
4. Sawasaki T, Kamura N, Matsunaga S, Saeki M, Tsuchimochi M et al (2008) Arabidopsis HY5 protein functions as a DNA-binding tag for purification and functional immobilization of proteins on agarose/DNA microplate. FEBS Lett 582:221–2280
5. Masaoka T, Nishi M, Ryc A, Endo Y, Sawasaki T (2008) The wheat germ cell-free based

screening of protein substrates of calcium/calmodulin-dependent protein kinase II delta. FEBS Lett 582:1795–1801

6. Nemoto K, Takemori N, Seki M, Shinozaki K, Sawasaki T (2015) Members of the plant CRK-superfamily are capable of trans-/auto-phosphorylation of tyrosine residues. J Biol Chem 290:16665–16677
7. Nemoto K, Seto T, Takahashi H, Nozawa A, Seki M et al (2011) Autophosphorylation profiling of Arabidopsis protein kinases using the cell-free system. Phytochemistry 72: 1136–1144
8. Eglen RM, Reisine T, Roby P, Rouleau N, Illy C et al (2008) The use of AlphaScreen technology in HTS: current status. Curr Chem Genomics 1:2–10
9. Takahashi H, Nozawa A, Seki M, Shinozaki K, Endo Y et al (2009) A simple and high-sensitivity method for analysis of ubiquitination and polyubiquitination based on wheat cell-free protein synthesis. BMC Plant Biol 9:39
10. Tadokoro D, Takahama S, Shimizu K, Hayashi S, Endo Y et al (2010) Characterization of a caspase-3-substrate kinome using an N- and C-terminally tagged protein kinase library produced by a cell-free system. Cell Death Dis 1:e89
11. Akagi T, Shimizu K, Takahama S, Iwasaki T, Sakamaki K et al (2011) Caspase-8 cleavage of the interleukin-21 (IL-21) receptor is a negative feedback regulator of IL-21 signaling. FEBS Lett 585:1835–1840
12. Takahashi H, Uematsu A, Yamanaka S, Imamura M, Nakajima T et al (2016) Establishment of a wheat cell-free synthesized protein array containing 250 human and mouse E3 ubiquitin ligases to identify novel interaction between E3 ligases and substrate proteins. PLoS One 11:e0156718

Chapter 12

Investigation of Autophosphorylation Sites of Plant Receptor Kinases and Phosphorylation of Interacting Partners

Matthew R. Meyer, Shweta Shah, and A. Gururaj Rao

Abstract

The optimal kinase activity of plant receptor-like kinases (RLKs) is often regulated by autophosphorylation of specific sites. Many of these phosphorylated residues then serve as recruiting sites for downstream interacting proteins. Therefore, identification of the phosphosites can be an important first step in delineating the signaling network. This chapter describes a protocol for identification of phosphorylated residues by mass spectrometry as well as a protocol to determine if an interacting partner can be phosphorylated in vitro.

Key words ACR4, Autophosphorylation, Mass spectrometry, LC-MS, Posttranslational modification, Phosphopeptide, Receptor-like kinases, Protein-protein interaction

1 Introduction

Protein phosphorylation is a key posttranslational modification (PTM) that occurs in both eukaryotes and prokaryotes and is important to many critical cellular activities. In plants, protein phosphorylation is primarily achieved by RLKs [1, 2]. In *Arabidopsis*, for example, more than 600 RLK genes have been identified [2]. Both in vivo and in vitro studies provide clear evidence of autophosphorylation in a variety of plant RLKs [3–5]. Evidence also indicates that phosphorylation of residues within the activation segment plays critical regulatory roles in the activation of many kinases [6, 7]. Importantly, phosphorylated residues also serve as recruitment sites for downstream interacting proteins and subsequent activation of signaling pathways that affect overall plant growth and development processes [8]. Identification of phosphorylation sites is an essential step in the biochemical characterization of the protein and delineating the signaling network.

Reidunn Birgitta Aalen (ed.), *Plant Receptor Kinases: Methods and Protocols*, Methods in Molecular Biology, vol. 1621, DOI 10.1007/978-1-4939-7063-6_12, © Springer Science+Business Media LLC 2017

2 Materials

2.1 In Vitro Phosphorylation Assay

1. Kinase assay buffer (5×): 100 mM Bis-Tris pH 7.2, 125 mM NaCl, 25 mM $MnCl_2$, 5 mM DTT.
2. Purified kinase solution (1 mg/mL).
3. 1 mM ATP solution.
4. 100% (w/v) TCA solution.
5. Ice cold acetone.
6. Microcentrifuge.

2.2 Phosphopeptide Enrichment

1. Denaturing buffer: 4 M urea, 50 mM NH_4HCO_3, pH 7.8.
2. 50 mM NH_4HCO_3, pH 7.8.
3. 250 mM TCEP.
4. 550 mM iodoacetamide, prepared fresh.
5. Sequencing grade modified trypsin.
6. 10% Trifluoroacetic acid (TFA).
7. C18 reverse phase column.
8. 90% acetonitrile (ACN), 0.1% formic acid (FA)
9. PHOS-Select™ Spin Column kit.
10. 60% ACN, 0.1% FA.
11. 1% ACN/0.1% FA.
12. 1.7 mL microcentrifuge tubes.
13. 0.5 mL microcentrifuge tubes.
14. Autosampler vials.

2.3 LC-MS Analysis of Phosphopeptides

1. Autosampler.
2. Nano-liquid chromatography instrument.
3. Mass spectrometer.
4. Analytical C18 reverse phase column.
5. 0.1% FA/99.9% H_2O.
6. 0.1% FA/99.9% ACN.
7. Database search engine (e.g., MASCOT).

2.4 Site-Directed Mutagenesis of Kinase Domains

1. DNA oligonucleotide primers.
2. QuikChange® Multi Site-Directed Mutagenesis kit.
3. Template plasmid containing coding sequence of the protein kinase.
4. PCR tubes.

2.5 In Vitro Phosphorylation of Putative Substrate Proteins by Receptor-like Kinase

1. Kinase assay buffer (5×): 100 mM Bis-Tris pH 7.2, 125 mM NaCl, 25 mM $MnCl_2$, 5 mM DTT.
2. [γ^{32}P]ATP (6000 Ci/mmol).
3. Phosphorimaging screen.
4. Scintillation counter.
5. LSC (liquid scintillation cocktail).

3 Methods

3.1 In Vitro Autophosphorylation

1. Prepare enough 5× kinase buffer required for the volume of reaction being performed. If necessary, the kinase buffer composition may also have to be optimized for each kinase by varying buffer conditions [4].
2. Add 40 μL of 5× kinase buffer to a 1.7 mL microfuge tube. Dilute to 100 μL with deionized water. Add 100 μL of purified kinase (concentration 1 μg/μL); the total reaction volume will be 200 μL.
3. Initiate the reaction by adding 2.5 μL of 1 mM ATP (final concentration is 25 μM).
4. Incubate the reaction at 25 °C for 60 min (*see* **Note 1**).
5. Terminate the reaction by 50 μL of TCA solution. Vortex to mix.
6. Incubate the reaction on ice for 30 min.
7. Spin sample at maximum speed for 15 min in a refrigerated microcentrifuge at 4 °C.
8. Aspirate the supernatant and discard.
9. Wash the pellet with 0.5 mL of ice cold acetone. Vortex briefly.
10. Spin sample at maximum speed for 15 min in a refrigerated microcentrifuge at 4 °C.
11. Aspirate the supernatant and discard.
12. Allow the pellet to air dry. Store at −80 °C until needed.

3.2 Sample Digestion and Phosphopeptide Enrichment

1. Resuspend the protein pellet containing autophosphorylated kinase in a 1.7 mL microfuge tube with 100 μL denaturing buffer. Vortex to mix.
2. Reduce the solubilized protein with 10 mM TCEP by adding 4.2 μL of 250 mM TCEP. Mix gently.
3. Incubate the reaction for 30 min at 30 °C.
4. Add 4.0 μL of 550 mM iodoacetamide (20 mM final concentration). Mix gently.
5. Incubate at room temperature for 30 min in the dark (*see* **Note 2**).

6. Add 3 volumes of 50 mM NH_4HCO_3, pH 7.8 to dilute the urea concentration below 1 M.
7. Add 2 μg of sequencing grade modified trypsin (1:50 w/w). Mix gently.
8. Incubate the reaction overnight at 37 °C.
9. Add 44 μL of 10% TFA to acidify the reaction to 1% TFA.
10. Desalt the peptides by solid phase extraction on a C18 reverse phase column.
11. Elute the peptides in a 0.5 mL tube with 90% ACN, 0.1% FA.
12. Dry down the eluate in a vacuum centrifuge.
13. Enrich phosphopeptides by immobilized metal affinity chromatography using the PHOS-Select™ Spin Column kit according to the kit instructions.
14. Elute phosphopeptides in a 0.5 mL tube with 60% ACN/0.1% FA.
15. Dry down the eluate in a vacuum centrifuge.
16. Resuspend the phosphopeptides in 25 μL of 1% ACN/0.1% FA and transfer to an autosampler vial for LC-MS analysis.

3.3 LC-MS Analysis of Phosphopeptides

LC-MS is a powerful and widespread technique utilized to identify post-translational modifications (PTMs; i.e., phosphorylation) of proteins. Generally, peptides go through a one-dimensional separation by a reverse phase chromatography succeeded by electrospray ionization at the mass spectrometer source. Peptides enter the mass spectrometer and undergo fragmentation followed by acquisition of MS/MS spectra. Peptides and PTMs are subsequently identified from matching MS/MS spectral information to a protein sequence database.

1. Inject 5 μL of enriched phosphopeptide sample on the LC-MS system.
2. Separate the phosphopeptides by reverse phase chromatography on a C18 analytical column with gradient conditions optimized for best elution profile and peptide resolution.
3. Continually collect MS/MS spectra of phosphopeptides using a mass spectrometer set in data-dependent acquisition mode (*see* **Note 3**).
4. Search the resulting MS/MS data against the protein kinase sequence allowing for variable modifications of phosphorylation at Ser, Thr, and Tyr residues and fixed modifications of carbamidomethylation at Cys residues.
5. A set of diagnostic, site-discriminating ions are used to verify phosphorylation site localization within a peptide of interest (Table 1) [4]. Manually verify the phosphopeptide identifications by comparing the MS/MS spectral information of each phosphopeptide with the theoretical exact mass of the phosphopeptide precursor and fragment ions (Fig. 1) (*see* **Note 4**) [4].

Table 1
Site-discriminating ions for phosphorylated residues (in *red*) of identified phosphopeptides

Peptide	b ion	Phosphorylated b ion	y ion	Phosphorylated y ion
[471]SSKDSAFTK[479]	b4 (+1, 0.14)	b5 (+1, 27.6)	y4 (+1, 1.14)	y5 (+2, 1.62)
[474]DSAFTKDNGK[483]	b4 (−)	b5 (+1, 4.9)	y5 (+1, 10.7)	y6 (+2, 2.5)
[499]VFTYEELEK[507]	b2 (+1, 6.04)	b3 (+1, 5.77)	y6 (+1, 12.82)	y7 (+1, 22.62)
[508]AADGFKEESIVGK[520]	b8 (+1, 0.77)	b9 (+1, 1.07)	y4 (+1, 0.45)	y5 (+1, 0.54)
[508]AADGFKEESIVGKGSF SCVYK[528]	b14 (+2, 0.82)	b15 (+2, 3.92)	y6 (+1, 6.07)	y7 (+1, 1.85)
[533]DGTTVAVK[540]	b2 (+1, 1.54)	b3 (+1, 1.34)	y5 (+1, 097)	y6 (+2, 2.91)
[533]DGTTVAVK[540]	b3 (+1, 2.03)	b4 (−)	y4 (+1, 3.51)	y5 (+1, 1.79)
[549]QKNSNEFR[556]	b3 (+1, 0.31)	b4 (−)	y4 (+2, 2.50)	y5 (+2, 1.46)
[551]NSNEFRTELDLLSR[564]	b6 (+1, 0.71)	b7 (+1, 2.56)	y7 (+1, 3.33)	y8 (+1, 0.55)
[557]TELDLLSR[564]	b6 (+1, 3.16)	b7 (+1, 3.64)	y1 (−)	y2 (−)
[641]DIKSSNILIDEEHNAR[656]	b3 (+1, 2.82)	b4 (+1, 2.63)	y12 (+3, 8.75)	y13 (+2, 26.46)
[641]DIKSSNILIDEEHNAR[656]	b4 (+1, 3.62)	b5 (−)	y11 (+2, 9.37)	y12 (+2, 6.06)
[826]SGSENTEFR[834]	b5 (+1, 2.38)	b6 (+1, 0.42)	y3 (+1, 5.06)	y4 (+1, 5.44)
[835]GGSWITFPSVTSSQR[849]	b2 (−)	b3 (−)	y12 (+2, 5.39)	y13 (+2, 4.55)
[875]SLEEEIGPASPGQSLFL HHNF[895]	(−)	b1 (−)	y20 (+2, 0.15)	y21 (−)
[875]SLEEEIGPASPGQSLFL HHNF[895]	b13 (−)	b14 (+1, 0.32)	y7 (+1, 0.48)	y8 (−)

The percentage of maximum intensity for each site-discriminating ion in the averaged high-resolution CID spectrum are shown in parenthesis (supporting ion information is available for underlined peptide sequences). (Reproduced from Ref. 4 with permission from American Chemical Society)

3.4 Production of Kinase-Inactive Mutants

Site-directed mutagenesis of residues within a kinase domain is a useful technique for generating an inactive kinase domain or for interrogating residues that are important to kinase autophosphorylation. The protocol below describes an abridged version of the instructions described in the QuikChange® Multi Site-Directed Mutagenesis kit.

1. Determine the codons that correspond to the amino acid that requires a mutation.
2. For the mutant codon, select the codon combination for optimal translation in the expression system being utilized.
3. Synthesize mutagenic DNA oligonucleotides with encoded point mutations [4].
4. Prepare the double-stranded DNA template plasmid by standard plasmid miniprep procedures.

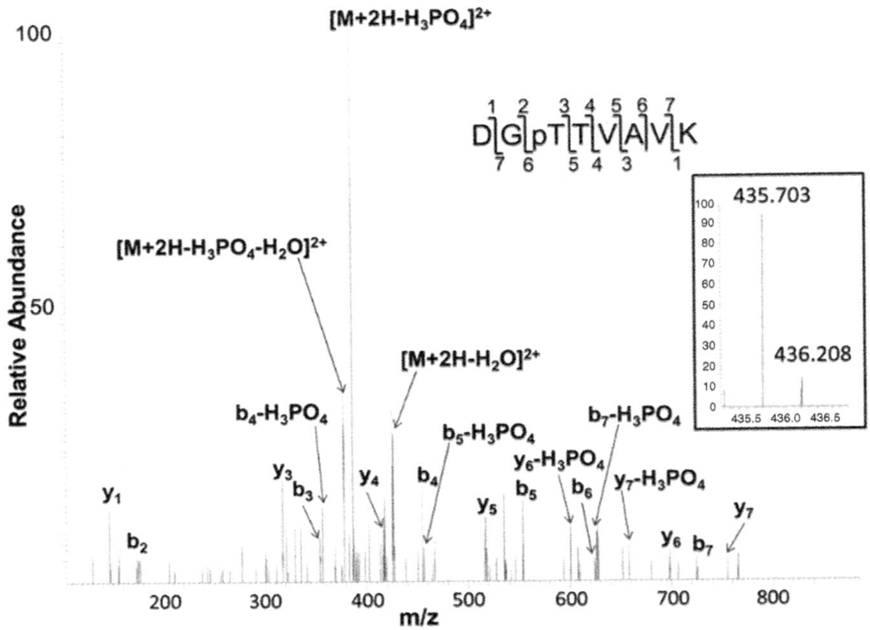

Fig. 1 Collision-induced low-resolution fragmentation spectrum of the phosphopeptide DGpTTVAVK (Mascot ion score 41), encompassing residues 533–540 of the *Arabidopsis* CRINKLY4 (ACR4) sequence. The presence of phosphorylated b3 (theor. *m/z* 354.1; obs'd 354.1) and y6 (theor. *m/z* 698.3; obs'd 698.2) ions, along with nonphosphorylated b2 (theor. *m/z* 173.1; obs'd 173.1) and y5 (theor. *m/z* 517.3; obs'd 517.3) ions, confirms Thr535 as the site of phosphorylation. The inset panel shows the MS spectrum of the doubly charged protonated molecular ion [M+2H]2+. Isotope clusters are separated by 0.505 Da, confirming the charge state of the peptide. (Reproduced from Ref. 4 with permission from American Chemical Society)

5. Prepare the mutant strand synthesis reactions according to the kit instructions (*see* **Note 5**).
6. Thermocycle the mutagenesis reaction for 30 iterations.
7. Cool the reaction on ice for 2 min.
8. Add 1 μL of Dpn I restriction enzyme to each reaction. Gently mix.
9. Incubate the reaction at 37 °C for 1 h.
10. Perform transformation with 45 μL of XL10-Gold ultracompetent cells and 1–10 μL of the newly created mutagenized plasmid (*see* **Note 6**).
11. Verify correct mutation that was inserted into the template plasmid by DNA sequencing.
12. Express and purify the protein kinase with established protocols.
13. Perform autophosphorylation assays as described in Subheading 3.1 to determine autophosphorylation activity of each mutant kinase (Fig. 2) [4].

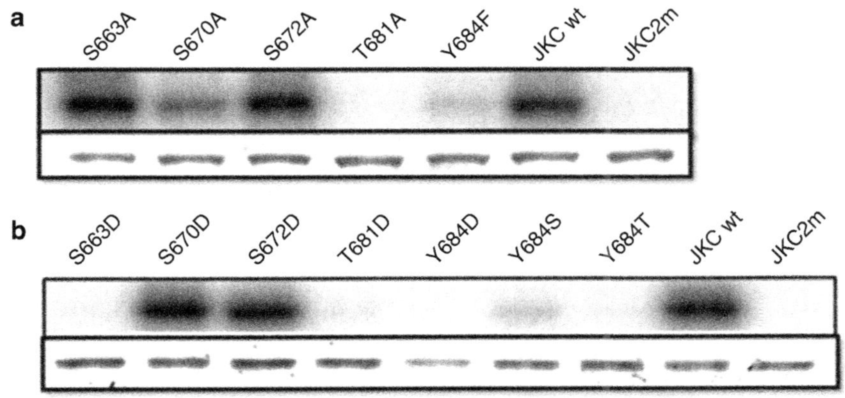

Fig. 2 Effect of mutating activation loop phosphorylation sites on kinase activity. (**a**) Mutations of S663, S670, S672, T681 to Ala and Y684 to Phe. (**b**) Mutations of S663, S670, S672, T681 to Asp and Y684 to Asp, Ser, and Thr. Each mutant was purified and subjected to in vitro kinase assay as described in Subheadings 2 and 3. Equal amounts of recombinant proteins were loaded and separated by SDS-PAGE followed by autoradiography. In (**a**) and (**b**), the *top panel* shows the autoradiogram and the *bottom panel* the corresponding Coomassie Blue-stained gel. Wild-type JKC and the mutant, JKC2m, were used as controls. (Reproduced from Ref. 4 with permission from American Chemical Society)

3.5 In Vitro Phosphorylation of Putative Substrate Proteins by Receptor-like Kinase

Protein-protein interactions and complexes mediated through the kinase domain are known to play critical roles in downstream signaling. Kinase-interacting partners can be identified by a variety of techniques and the proteins isolated by pull-down experiments. Often, the interacting protein can serve as a substrate for the kinase, leading to phosphorylation of specific sites (Fig. 3). The following procedure can be used to test the phosphorylation of a putative substrate.

1. Make sufficient 5× kinase buffer according to the number of reactions to be performed.
2. Dilute the [γ^{32}P]ATP (6000 Ci/mmol) with 20 mM Bis-Tris pH 7.2 so that final concentration of the stock is 1 μCi/μL. Make sure proper safety precautions are exercised while working with [γ^{32}P]ATP.
3. Add 4 μL of 5× kinase buffer, 0.5 μL of 1 mM ATP (final concentration is 25 μM), and 2 μL of diluted [γ^{32}P]ATP (final amount is 2 μCi), depending upon protein concentration; adjust the final volume of reaction with deionized water so that the total reaction volume is 20 μL.
4. Start the reaction by adding 1 μg of RLK and 1 μg of substrate protein or inactive kinase mutant (prepare as described in Subheading 3.4).
5. Incubate the reaction at 25 °C for 60 min (*see* **Note 1**).
6. Stop the reaction by adding 5 μL of 4× Laemmli buffer and heating the reaction mixture to 95 °C for 5 min.

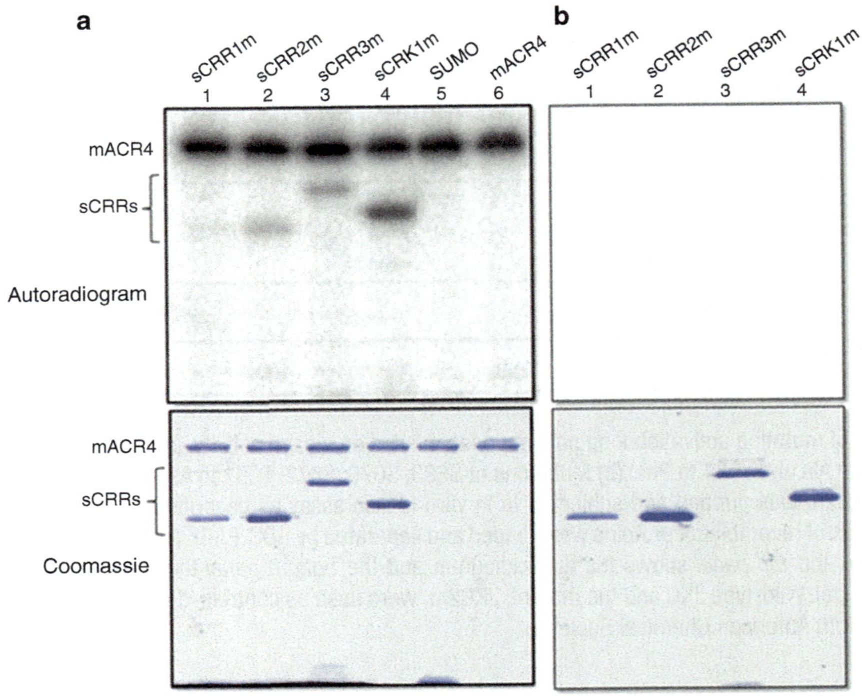

Fig. 3 *Arabidopsis* CRINKLY4 (ACR4) kinase domain can phosphorylate the intracellular domains of the *Arabidopsis* CRINKLY4-related (AtCRRs) proteins. Active MBP-ACR4 was incubated with the inactive SUMO-CRRs in an in vitro kinase assay. (**a**) *Upper panel* shows an autoradiogram demonstrating ACR4 can phosphorylate the ICDs of the inactive CRRs. ACR4 *cannot* phosphorylate the SUMO tag. *Lower panel* shows the corresponding Coomassie stained gel. (**b**) *Upper panel* depicts an autoradiogram showing the inactivity of the sCRR mutants in which their respective catalytic Asp residues have been mutated to Ala. *Lower panel* is the corresponding Coomassie stained gel showing the presence of each protein. (Reproduced from Ref. 9 with permission from PloS One)

7. Resolve the protein on SDS-PAGE (*see* **Note 7**). Stain the gels with Coomassie Brilliant Blue R-250.
8. Wrap the gel in saran wrap. Expose the gels onto phosphorimaging screen for 30 min or gels can also be exposed to autoradiogram film (*see* **Note 8**). Remove the gels from the screen in dark to avoid exposure of light.
9. Scan the phosphorimaging screen on a Typhoon scanner.
10. Quantitative determination of kinase activity is measured by scintillation counting.
11. Cut each protein band from the gels and also an unphosphorylated protein band for a background reading.
12. Place each band in separate scintillation vials containing 5 mL of LSC liquid.
13. Prepare an ATP standard curve by using serially diluted radioactive ATP. Dilute the ATP 10×, 100×, 1000×, and 10,000×

from the stock solution. Spot 10 μL of each standard dilution on separate 10 mm × 10 mm size Whatman filter paper and place in scintillation vials containing 5 mL of LSC solution (*see* **Note 9**).

14. Quantify the radioactive counts for each sample and each standard vial with a scintillation counter.
15. Back calculate the amount of μCi of ATP used for each standard. Plot the standard curve against μCi of ATP used versus counts per minute.
16. From the standard curve, calculate the amount of ATP incorporated for each reaction. Convert the μCi into fmol (1 fmol = 0.006 μCi) used. Calculate kinase-specific activity in terms of amount of fmol of ^{32}P incorporated per pmol of protein per min.

4 Notes

1. Some kinases may be temperature sensitive and may require the reaction to be performed at lower temperatures. The time of incubation can also be optimized for a particular protein.
2. Iodoacetamide is unstable and light sensitive.
3. Eluted phosphopeptides are ionized at the source and fragmented within the mass spectrometer using a mass spectrometry method optimized for maximum sensitivity and acquisition of good quality MS/MS spectra for database searching.
4. The exact masses of the phosphopeptide and fragmentation ions can be calculated using the MS-Product utility within Protein Prospector (http://prospector.ucsf.edu).
5. The ratio of the reaction components may have to be optimized for template length and number of mutagenic primers used.
6. The volume of plasmid used to transform the XL10-Gold ultracompetent cells may need to be optimized to produce an adequate number of colony-forming units for verification of correct mutation.
7. Stop running the gel when the dye front is 1–2 cm away from the bottom front. This will prevent contamination of the gel running into the tank by unincorporated radioactive ATP. Cut the gel above the dye front and dispose it according to radioactive waste disposal protocols.
8. Optimal exposure time may vary between 15 min and 1 h depending on kinase activity.
9. Standard curve needs to be prepared fresh since the radioactivity changes with storage.

References

1. De Smet I, Voss U, Jurgens G, Beeckman T (2009) Receptor-like kinases shape the plant. Nat Cell Biol 11:1166–1173
2. Shiu SH, Bleecker AB (2001) Plant receptor-like kinase gene family: diversity, function, and signaling. Sci STKE 2001:re22
3. Oh MH, Wu X, Clouse SD, Huber SC (2011) Functional importance of BAK1 tyrosine phosphorylation in vivo. Plant Signal Behav 6:400–405
4. Meyer MR, Lichti CF, Townsend RR, Rao AG (2011) Identification of in vitro autophosphorylation sites and effects of phosphorylation on the Arabidopsis CRINKLY4 (ACR4) receptor-like kinase intracellular domain: insights into conformation, oligomerization, and activity. Biochemist 50:2170–2186
5. Liu GZ, Pi LY, Walker JC, Ronald PC, Song WY (2002) Biochemical characterization of the kinase domain of the rice disease resistance receptor-like kinase XA21. J Biol Chem 277:20264–20269
6. Johnson LN, Lewis RJ (2001) Structural basis for control by phosphorylation. Chem Rev 101:2209–2242
7. Burza AM, Pekala I, Sikora J, Siedlecki P, Malagocki P et al (2006) *Nicotiana tabacum* osmotic stress-activated kinase is regulated by phosphorylation on Ser-154 and Ser-158 in the kinase activation loop. J Biol Chem 281: 34299–34311
8. Schulze WX, Deng L, Mann M (2005) Phosphotyrosine interactome of the ErbB-receptor kinase family. Mol Syst Biol 1:42–55
9. Meyer MR, Shah S, Zang J, Rohrs H, Rao AG (2015) Evidence for intermolecular interactions between the intracellular domains of the *Arabidopsis* receptor like kinase ACR4, its homologs and the WOX5 transcription factor. PLoS One 10(3):e0118861

Chapter 13

In Vitro Assessment of Guanylyl Cyclase Activity of Plant Receptor Kinases

Misjudeen Raji and Chris Gehring

Abstract

Cyclic nucleotides such as 3′,5′-cyclic adenosine monophosphate (cAMP) and 3′,5′-cyclic guanosine monophosphate (cGMP) are increasingly recognized as key signaling molecules in plants, and a growing number of plant mononucleotide cyclases, both adenylate cyclases (ACs) and guanylate cyclases (GCs), have been reported. Catalytically active cytosolic GC domains have been shown to be part of many plant receptor kinases and hence directly linked to plant signaling and downstream cellular responses. Here we detail, firstly, methods to identify and express essential functional GC domains of receptor kinases, and secondly, we describe mass spectrometric methods to quantify cGMP generated by recombinant GCs from receptor kinases in vitro.

Key words Cyclic nucleotides, Nucleotide cyclase, Guanylate cyclase, cGMP, HPLC, Tandem mass spectrometry

1 Introduction

Since the identification and characterization of the first GC in higher plants [1], there has been a renewed interest in mononucleotide cyclases and cyclic nucleotide signaling in plants (for review *see* [2, 3]) not least because many physiological processes are directly or indirectly modulated by 3′,5′-cyclic guanosine monophosphate (cGMP). These processes include cGMP-dependent changes in stomatal aperture [4], the transcriptional regulation of ozone responsive genes [5], the regulation of K^+ transport [6, 7], and cyclic nucleotide-gated channels (CNGCs) [8] as well as cGMP-dependent phosphorylation [9].

It has turned out that many of the predicted and experimentally tested plant GCs are in fact receptor kinases. In *Arabidopsis thaliana*, the characterized receptor kinases include a wall associated kinase-like protein (AtWAKL-10) [10], the brassinosteroid receptor [11], the phytosulfokine receptor (PSKR1) [12], the receptor for the plant elicitor peptide 1 (AtPepR1) [13], and the

Reidunn Birgitta Aalen (ed.), *Plant Receptor Kinases: Methods and Protocols*, Methods in Molecular Biology, vol. 1621, DOI 10.1007/978-1-4939-7063-6_13,

receptor for the plant natriuretic peptide (AtPNP-R1) [14]. Furthermore, it has also recently been demonstrated that in the PSKR1, Ca^{2+} acts as a switch: switching on GC activity and switching off kinase activity [15–17].

Given that the single-celled green alga *Chlamydomonas reinhardtii* harbors >90 annotated nucleotide cyclases, GC and ACs, that come in >20 different domain combinations including receptor kinases [18], we predicted that in higher plants too, many more plant receptor kinases with functional GC catalytic centers will be discovered [19] and tested in vitro and in vivo.

Taken together, the continuing expansion of the field of cGMP-dependent signaling requires the application of reliable methods for the assessment of catalytic functions and mechanisms of cyclic nucleotide cyclases in general and receptor-coupled GCs in particular.

Mass spectrometry-based methods for the determination of cyclic nucleotides have been widely reported in the literature, with different types of mass analyzers coupled to high-performance liquid chromatography. Methods based on triple quadrupole mass analyzers are most extensively reported [20–24] as they offer high sensitivity coupled with multi-analyte detection capability. Ion trap mass analyzers have also been utilized [25, 26], and they offer the advantage of low cost and the capacity to perform MS^n fragmentations that can yield more structural information. The method reported herein is based on an Orbitrap mass analyzer but is adaptable to any ion trap type instrument.

2 Materials

The minimum required instrumentation includes a HPLC (e.g., Accela Quaternary Pump, Thermo Scientific) (*see* **Note 1**) and a mass spectrometer capable of multistage MS analysis, such as the LTQ-Orbitrap Velos (Thermo Scientific) (*see* **Note 2**).

Solutions for the GC in vitro assay, sample and standard preparation, and subsequent LC-MS/MS analysis must be made up in at least HPLC-grade or, better still, HPLC-MS-grade solvents (e.g., water, acetonitrile).

2.1 Obtaining Recombinant or Synthetic GCs

Identification of candidate GCs, cloning and expressing of candidate GCs, and the isolation of recombinant GC have been described extensively ([3] and references therein). Given that receptor kinases containing GC domains are often large transmembrane proteins, it may be desirable to clone and express only the (cytosolic) GC domain (*see* **Note 3**). However, using the cytosolic domain only will not allow to study ligand-dependent effects on GC activity. Furthermore, there are numerous companies that offer peptide synthesis services, and it may be expedient to have a synthetic candidate GC domain synthesized for functional testing.

2.2 GC In Vitro Assay and Preparation of cGMP Sample

All reagents should be prepared and stored in sterile plastic ware:

1. GC protein (either recombinant or synthesized): the protein concentration needs to be of determined accurately (*see* **Note 4**).
2. 50 mM Tris (*Tris*(hydroxymethyl)aminomethane) pH 7.5: For a 50 mL solution, dissolve 0.3 g Tris in 30 mL of deionize (di) H_2O water and mix the solution on a magnetic stirrer. Adjust the pH of the solution to 7.5 and then make up with diH_2O to a final volume of 50 mL.
3. 10 mM GTP stock solution: Weigh out 5.23 mg of GTP into a 1.5 mL microcentrifuge tube and add 1 mL of diH_2O, and mix by vortexing till particles have dissolved completely.
4. cGMP stock solution: Weigh out 1 mg of cGMP and dissolve in 1 mL of diH2O (or better still HPLC-grade H_2O) and store at 4 °C until further use.
5. 0.5 M stock solution of $MgCl_2$: Dissolve 2.54 g $MgCl_2 \bullet 6H_2O$ (if you are using the hexahydrate form) into 25 mL of water in a 50 mL sterile plastic tube (e.g., FalconTM) and mix by vortexing.
6. 0.5 M stock solution of $MnCl_2$: Dissolve 4.95 g $MnCl_2 \bullet 6H_2O$ (if you are using the hexahydrate form) into 25 mL of diH_2O in a 50 mL plastic tube and mix by vortexing.
7. Ca^{2+} stock solutions of various concentrations (*see* **Note 5**).
8. Optional: 50 mM stock solution of the phosphodiesterase inhibitor isobutylmethylxanthine (IBMX): Weigh out 22.2 mg of IBMX into a 2 mL Eppendorf tube and add 2 mL of dimethyl sulfoxide (DMSO), and then mix the solution by vortexing. IBMX is a potent nonspecific inhibitor of phosphodiesterases and is not essential in in vitro studies but highly advisable when cGMP is extracted form cells as part of an in vivo assay.
9. Reaction mix for the GC assay: 50 mM Tris–HCl (pH 7.4) in the presence of 10 mM GTP or 10 mM ATP, 5 mM Mg^{2+}, or 5 mM Mn^{2+}.

2.3 Chromatography

1. Chromatography is performed with an Accela HPLC system (Thermo Scientific) equipped with a quaternary pump (*see* **Note 1**) and an autosampler-column compartment module. The system is coupled to a mass spectrometer for analyte detection.
2. The mobile phase used for separation comprises a 60/40 isocratic mixture of 10 mM ammonium acetate in water and LC-MS-grade acetonitrile. High-quality water, e.g., purified using the Milli-Q system, should be used for the preparation of the ammonium acetate buffer (*see* **Note 6**).
3. The column to use is a Sepax SFC Cyano column (150 mm × 4.6 mm, 5 μm) (*see* **Note 7**).

2.4 Spectrometry

1. The method has been tested using an LTQ-Orbitrap Velos mass spectrometer system fitted with an Ion Max II electrospray ionization source (*see* **Note 2**). The MS instrument comprises a frontend linear ion trap and a backend Orbitrap mass analyzer.
2. Nitrogen gas from liquid nitrogen Dewar is used as sheath gas as well as auxiliary gas (*see* Table 1).
3. Helium is used as damping gas in the ion trap as well as the collision gas during MS/MS analysis (*see* **Note 8**).
4. cGMP calibration standards.

Table 1
Instrument information and analytical method summary

Instrumentation	
HPLC	Accela HPLC (Thermo)
Column	Sepax SFC Cyano (Sepax); 150 × 4.6 mm, 5 μm
Mass spectrometer	LTQ-Orbitrap Velos
HPLC parameters	
HPLC Mobile Phase	Isocratic mix of 10 mM ammonium acetate and acetonitrile at 60%/40% ratio
Injection Volume	10 μL
Flow Rate	600 μL/min
Column Temperature	Ambient
MS parameters	
Ionization mode	Positive ESI
Source parameters	Spray voltage = 3.8 kV Vaporizer temperature = 310 °C Sheath gas = 30 Auxiliary gas = 10 Ion transfer capillary temperature = 270 °C S-Lens voltage = 60
Fragmentation type	High Collision Dissociation (HCD)
Precursor analyzer	Linear trap
Fragment analyzer	Orbitrap
Normalized collision energy	30%
Activation time	0.1 ms
Resolution	30,000
Precursor ion	346.056
Product ion	152.056

2.5 Data Analyses Software

Instrument control and data analysis can be performed using the Excalibur software (Thermo Scientific, version 2.2).

3 Methods

3.1 In Vitro GC Assay

1. Determine the number of different reactions you will have carry out, and then prepare the standard reaction mix for the GC assay (*see* **Note 9**).
2. Start the reaction by adding 1 μg either the purified recombinant or synthetic peptide in ≤5 μL of diH_2O to 50 μL of reaction mix.
3. Incubate for 20 min at room temperature (20 °C) (*see* **Note 10**).
4. Terminate the reactions by boiling for 3 min in the tube, and then cool the samples down on ice for 2 min, followed by centrifugation at 2300 × *g* for 3 min.
5. Discard the pellet and dilute the supernatants ≥5-fold in diH_2O water. At this stage, the sample is ready for LC-MS/MS analysis or can be dried and stored.

3.2 Separation and Quantification of cGMP

1. Separation of cGMP can be carried out using an isocratic mobile phase mixture of 10 mM ammonium acetate and acetonitrile at 60%/40% ratio. The flow rate is typically 600 μL/min (*see* **Note 11**).
2. Calibration: Prepare an intermediate standard solution of 1000 pg/μL from the cGMP stock by dilution, and use this for preparing the calibration standards of 0.5, 1, 2, 5, 10, 20, 50, and 100 pg/μL.
3. Preparation of extracted samples for LC-MS/MS analysis: If the extracted samples were dried, they need to be reconstituted in ≥200 μL of HPLC-grade H_2O by vortexing for ≥10 s and then centrifuged for 10 min at 4 °C at >20,000 × *g*. Sample can be spiked with the internal standards if applicable.
4. Inject 10 μL of each sample and internal calibration standard for analysis. Air bubbles must be avoided.
5. MS analysis is performed in the positive electrospray ionization mode. The MS source parameters and other instrument parameters are summarized in Table 1 (*see* **Note 12**).
6. The detection is based on selected reaction monitoring of the cGMP molecule by fragmenting its precursor ion at m/z 346.05 and yielding a product ion at 152.05 (*see* **Notes 13–15**).
7. The standard calibration curve will be based on the peak areas of each calibration concentration using the extracted ion chromatogram of product ion m/z 152.05. The lowest detectable amount is typically 0.1 pg/μL (*see* **Note 16**).
8. Quantitation is based on the chromatographic peak areas of the samples using the extracted ion chromatogram of product ion m/z 152.05 (*see* Fig. 1).

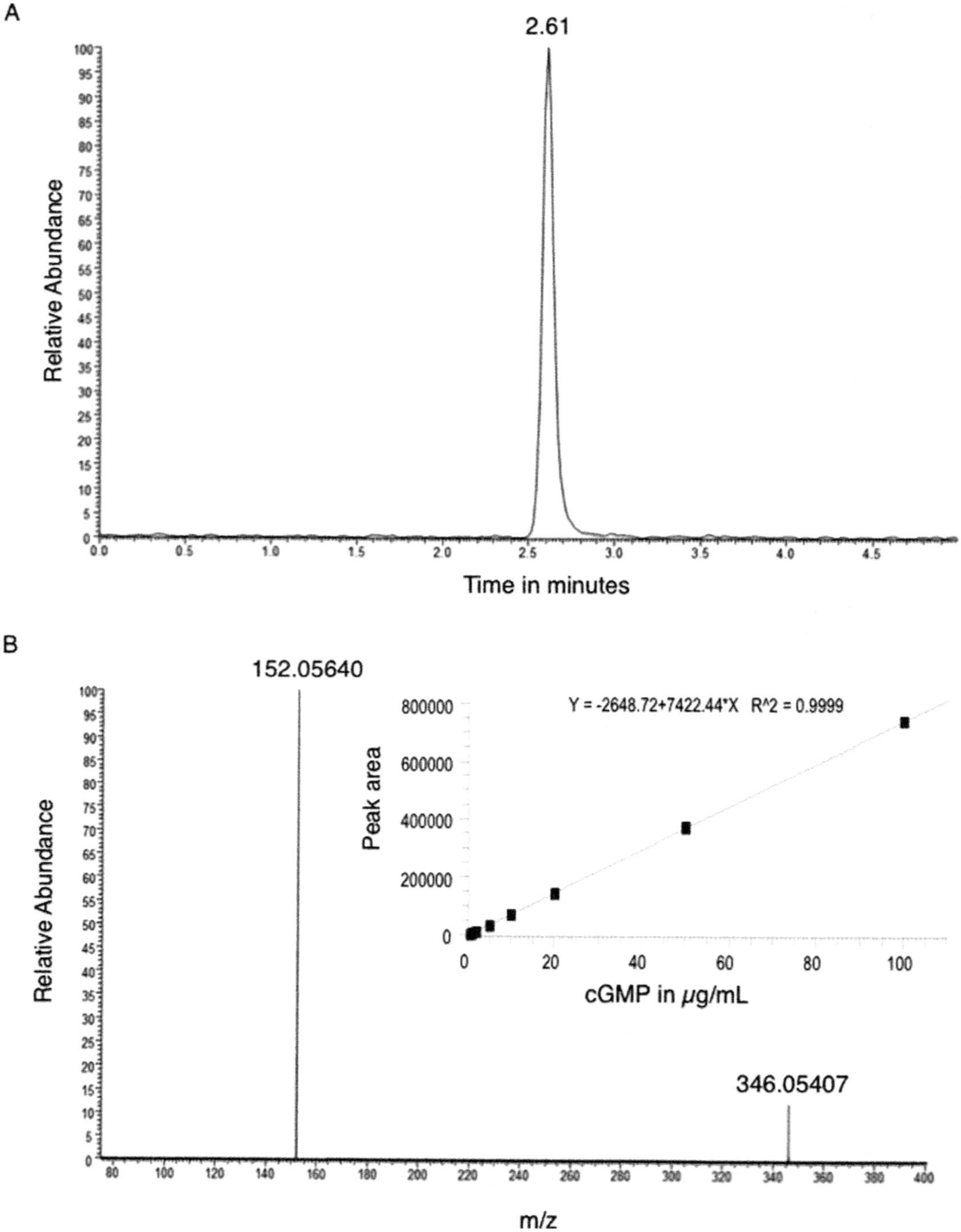

Fig. 1 Abundance of cGMP. (**A**) Example extracted ion chromatogram of cGMP product ion with *m/z* 152.05 at 20 pg/μL. The retention time of cGMP under the afore-described mobile phase and column conditions is 2.61 min. (**B**) Mass spectrum of cGMP showing the transition of parent ion at *m/z* 346.05 $[M+H]^+$ to the product ion at *m/z* 152.05. Inset: Standard Calibration curve for cGMP between 0.1 and 100 pg/μL

4 Notes

1. Using a multichannel HPLC pump (such as binary or quaternary) allows flexibility of solvent change and gradient elution during method development and optimization. However, an isocratic pump may also be used in this method since the mobile phase composition is the same throughout the run.
2. Other mass spectrometers such as ion traps and triple quadrupoles are also suitable. Simultaneous determination and quantitation of multiple compounds are possible when using the MRM mode of analysis in triple quadrupoles even when the compounds are not chromatographically well separated. In ion traps however, MS/MS mode is best suited for simultaneous quantitation of multiple compounds if the compounds are already chromatographically separated.
3. For GC in vitro assays, it is often sufficient to clone the GC domain of plant receptor kinases only which typically spans about 100 amino acids with 50 amino acids on each side of the catalytic center (e.g., [11]). If the ligand dependence of a GC will be tested, the entire plant receptor kinase, including the ligand-binding domain, needs to be expressed (e.g., [14]).
4. Accurate determination of the protein concentration of the extracted recombinant or synthesized protein is essential for the quantification of GC activities. The determination can be done with a Bio-Rad protein assay kit that is based on the Bradford method.
5. For all experiments, Ca^{2+} concentrations should be kept at a physiological range of ≈10 nM. Free calcium concentrations can be determined using the MAXCHELATOR program (http://maxchelator.stanford.edu/ (last accessed March 2016)) taking into account the temperature, pH, and ionic strength of the calcium buffer.
6. High-quality water (i.e., HPLC-grade water) must be used for the mobile phase buffer as well as for the sample preparation. Excessive background ions can compromise the detection sensitivity of the instrument.
7. Other columns such as C18 and C8 may be used. However, better retention was observed for cGMP under the mobile phase composition used in this method with the Cyano column compared to C18 and C8. This is very important when dealing with multiple compounds that are analogous to cGMP (such as GMP and cAMP).
8. Triple quadrupole mass spectrometers typically use argon as collision gas; however, nitrogen may also be used where cost so demands.

9. Addition of Ca^{2+} to the standard reaction mix is required if Ca^{2+}-dependence of the CG is studied.
10. For kinetic studies, a time series will have to be chosen and the times typically span from 30 s to 30 min.
11. The organic component of the mobile phase can have a significant effect on the elution and must be carefully selected and optimized. Methanol and acetonitrile are the most common organic modifiers in LC-MS mobile phases. Acetonitrile was preferred in this method due to its strong elution power compared to methanol. Acetonitrile reduces run times and has the capacity to remove other components in the sample matrix from the column more efficiently. This is particularly important when dealing with complex matrices. It also enhances the analyte ionization efficiency during ESI-MS (though this is sometimes compound dependent) and generates less column back pressure during HPLC. It should be noted, however, that faster elution may sometimes compromise chromatographic resolution when multiple compounds are being monitored that elute closely on the column.
12. The instrument parameters listed in Table 1 may be different from those of other mass spectrometers (or similar mass spectrometers from other manufacturers for that matter). It is therefore essential to tune the instrument for optimal results following manufacturer's recommendations.
13. Depending on the type of mass spectrometer, the fragmentation modes available for MS/MS analysis may be different. During the optimization stage for this method, the high collision dissociation (HCD) mode (that is mainly a feature of Orbitrap instruments) was found to be more efficient yielding more ion counts and therefore better sensitivity as compared to the collision-induced dissociation (CID) on the linear ion trap of the same instrument.
14. The collision energy must be optimized for best sensitivity. In this method, we used a collision energy that allowed at least 85% conversion of precursor to product. Note that less energy is required to activate the molecule in order to achieve this fragmentation level in HCD mode compared to CID.
15. In other instruments such as triple quadrupole, it is necessary to also optimize the collision gas pressure for best result. Excessive pressure may lead to peak distortion, while insufficient pressure may lead too low ion yield for the product.
16. The LOD and LOQ for this method were estimated as 0.1 and 0.5 pg/μL, respectively. The need for extensive method validation is based on the intended use of the data and may not be required in cases where relative comparison of amounts is the sole purpose.

Acknowledgments

We are grateful to the KAUST Analytical Core Lab for supporting this project.

References

1. Ludidi N, Gehring C (2003) Identification of a novel protein with guanylyl cyclase activity in *Arabidopsis thaliana*. J Biol Chem 278:6490–6494
2. Gehring C (2010) Adenyl cyclases and cAMP in plant signaling – past and present. Cell Commun Signal 8:15
3. Marondedze C, Wong A, Thomas L, Irving H, Gehring C (2016) Cyclic nucleotide monophosphates in plants and plant signaling. Handb Exp Pharmacol (in press) doi:10.1007/164_2015_35
4. Pharmawati M, Billington T, Gehring CA (1998) Stomatal guard cell responses to kinetin and natriuretic peptides are cGMP-dependent. Cell Mol Life Sci 54:272–276
5. Ederli L, Meier S, Borgogni A, Reale L, Ferranti F, Gehring C, Pasqualini S (2008) cGMP in ozone and NO dependent responses. Plant Signal Behav 3:36–37
6. Hoshi T (1995) Regulation of voltage dependence of the KAT1 channel by intracellular factors. J Gen Physiol 105:309–328
7. Ordoñez NM, Marondedze C, Thomas L, Pasqualini S, Shabala L, Shabala S, Gehring C (2014) Cyclic mononucleotides modulate potassium and calcium flux responses to H_2O_2 in Arabidopsis roots. FEBS Lett 588:1008–1015
8. Zelman AK, Dawe A, Gehring C, Berkowitz GA (2012) Evolutionary and structural perspectives of plant cyclic nucleotide-gated cation channels. Front Plant Sci 3:95
9. Marondedze C, Groen AJ, Thomas L, Lilley KS, Gehring C (2016) A quantitative phosphoproteome analysis of cGMP-dependent cellular responses in *Arabidopsis thaliana*. Mol Plant 9:621–623
10. Meier S, Ruzvidzo O, Morse M, Donaldson L, Kwezi L, Gehring C (2010) The Arabidopsis wall associated kinase-like 10 gene encodes a functional guanylyl cyclase and is co-expressed with pathogen defense related genes. PLoS One 5:e8904
11. Kwezi L, Meier S, Mungur L, Ruzvidzo O, Irving H, Gehring C (2007) The *Arabidopsis thaliana* brassinosteroid receptor (AtBRI1) contains a domain that functions as a guanylyl cyclase *in vitro*. PLoS One 2:e449
12. Kwezi L, Ruzvidzo O, Wheeler JI, Govender K, Iacuone S, Thompson PE, Gehring C, Irving HR (2011) The phytosulfokine (PSK) receptor is capable of guanylate cyclase activity and enabling cyclic GMP-dependent signaling in plants. J Biol Chem 286:22580–22588
13. Qi Z, Verma R, Gehring C, Yamaguchi Y, Zhao Y, Ryan CA, Berkowitz GA (2010) Ca^{2+} signaling by plant *Arabidopsis thaliana* Pep peptides depends on AtPepR1, a receptor with guanylyl cyclase activity, and cGMP-activated Ca^{2+} channels. Proc Natl Acad Sci U S A 107:21193–21198
14. Turek I, Gehring C (2016) The plant natriuretic peptide receptor is a guanylyl cyclase and enables cGMP-dependent signaling. Plant Mol Biol 91(3):275–286
15. Muleya V, Wheeler JI, Irving HR (2013) Structural and functional characterization of receptor kinases with nucleotide cyclase activity. Methods Mol Biol 1016:175–194
16. Freihat L, Muleya V, Manallack DT, Wheeler JI, Irving HR (2014) Comparison of moonlighting guanylate cyclases: roles in signal direction? Biochem Soc Trans 42:1773–1779
17. Muleya V, Wheeler JI, Ruzvidzo O, Freihat L, Manallack DT, Gehring C, Irving HR (2014) Calcium is the switch in the moonlighting dual function of the ligand-activated receptor kinase phytosulfokine receptor 1. Cell Commun Signal 12:60
18. Meier S, Seoighe C, Kwezi L, Irving H, Gehring C (2007) Plant nucleotide cyclases: an increasingly complex and growing family. Plant Signal Behav 2:536–539
19. Wong A, Gehring C (2013) The *Arabidopsis thaliana* proteome harbors undiscovered multi-domain molecules with functional guanylyl cyclase catalytic centers. Cell Commun Signal 11:48
20. Van Damme T, Zhang Y, Lynen F, Sandra P (2012) Determination of cyclic guanosine- and cyclic adenosine monophosphate (cGMP and cAMP) in human plasma and animal tissues by solid phase extraction on silica and liquid

chromatography-triple quadrupole mass spectrometry. J Chromatogr B Analyt Technol Biomed Life Sci 909:14–21

21. Martens-Lobenhoffer J, Dautz C, Bode-Boger SM (2010) Improved method for the determination of cyclic guanosine monophosphate (cGMP) in human plasma by LC-MS/MS. J Chromatogr B Analyt Technol Biomed Life Sci 878:487–491
22. Zhang Y, Dufield D, Klover J, Li W, Szekely-Klepser G, Lepsy C, Sadagopan N (2009) Development and validation of an LC-MS/MS method for quantification of cyclic guanosine 3′,5′-monophosphate (cGMP) in clinical applications: a comparison with a EIA method. J Chromatogr B Analyt Technol Biomed Life Sci 877:513–520
23. Oeckl P, Ferger B (2012) Simultaneous LC-MS/MS analysis of the biomarkers cAMP and cGMP in plasma, CSF and brain tissue. J Neurosci Methods 203:338–343
24. Witters E, van Dongen W, Esmans EL, van Onckelen HA (1997) Ion-pair liquid chromatography-electrospray mass spectrometry for the analysis of cyclic nucleotides. J Chromatogr B Biomed Sci Appl 694:55–63
25. Jia X, Fontaine BM, Strobel F, Weinert EE (2014) A facile and sensitive method for quantification of cyclic nucleotide monophosphates in mammalian organs: basal levels of eight cNMPs and identification of 2′,3′-cIMP. Biomol Ther 4:1070–1092
26. Goutier W, Spaans PA, van der Neut MA, McCreary AC, Reinders JH (2010) Development and application of an LC-MS/MS method for measuring the effect of (partial) agonists on cAMP accumulation *in vitro*. J Neurosci Methods 188:24–31

Chapter 14

BiFC Assay to Detect Calmodulin Binding to Plant Receptor Kinases

Cornelia Fischer, Margret Sauter, and Petra Dietrich

Abstract

Plant receptor-like kinases (RLKs) are regulated at various levels including posttranscriptional modification and interaction with regulatory proteins. Calmodulin (CaM) is a calcium-sensing protein that was shown to bind to some RLKs such as the PHYTOSULFOKINE RECEPTOR1 (PSKR1). The CaM-binding site is embedded in subdomain VIa of the kinase domain. It is possible that many more of RLKs interact with CaM than previously described. To unequivocally confirm CaM binding, several methods exist. Bimolecular fluorescence complementation (BiFC) and pull-down assays have been successfully used to study CaM binding to PSKR1 and are described in this chapter (BiFC) and in Chapter 15 (pull down). The two methods are complementary. BiFC is useful to show localization and interaction of soluble as well as of membrane-bound proteins in planta.

Key words Agrobacterium-mediated transformation of *N. benthamiana*, BiFC, Confocal laser scanning microscope, Calmodulin, Peptide receptor kinase, Protein binding

1 Introduction

Plants possess a large number of receptor-like kinases (RLKs) but the function of only few has been uncovered, and their mode of regulation is for most RLKs not known. For RLKs, including the peptide receptor PSKR1 (PHYTOSULFOKINE RECEPTOR1), interaction with the co-receptor BAK1 (BRI1 associated kinase1) and with calmodulin (CaM) has been shown. The interaction domain for CaM corresponds to helix E of the kinase and is an amphipathic α-helix in kinase subdomain VIa [1]. CaM-binding sites are conserved and can be predicted with bioinformatic tools. Yet actual binding has to be confirmed experimentally. For PSKR1, binding to CaM in vivo was demonstrated by the bimolecular fluorescence complementation (BiFC) method [1, 2]. It allows the investigation of protein interactions in living cells and visualizes the subcellular localization of the interaction. BiFC studies are therefore well suited to investigate the interaction of membrane

Reidunn Birgitta Aalen (ed.), *Plant Receptor Kinases: Methods and Protocols*, Methods in Molecular Biology, vol. 1621,
DOI 10.1007/978-1-4939-7063-6_14, © Springer Science+Business Media LLC 2017

proteins. The method is based on the assembly of nonfluorescent halves of yellow fluorescent protein (YFP) or the modified version Venus to a functional fluorophore upon protein interaction and is therefore also called split-YFP or split-Venus, respectively. BiFC has been successfully adapted for assays in plant cells, mainly using *Nicotiana benthamiana* as the plant expression system and Venus as the fluorophore [3]. The N-terminal part of Venus (1–173, VN) and its C-terminal part (156–239, VC) are fused to different genes, in order to investigate the interaction of the corresponding fusion proteins (Fig. 1a).

BiFC assays revealed the interaction between the plasma membrane receptor kinase PSKR1 and the soluble CaMs in planta at the plasma membrane. A mutation of a conserved tryptophane in the CaM-binding domain to a serine (W831S) in helix E of the cytosolic kinase domain weakened its CaM-binding ability [1]. Since the expression of the mutated receptor PSKR1(W831S) alike the kinase-inactive receptor did not complement the growth defect of *pskr* null mutants, CaM binding was concluded to play an important role in PSKR1 function.

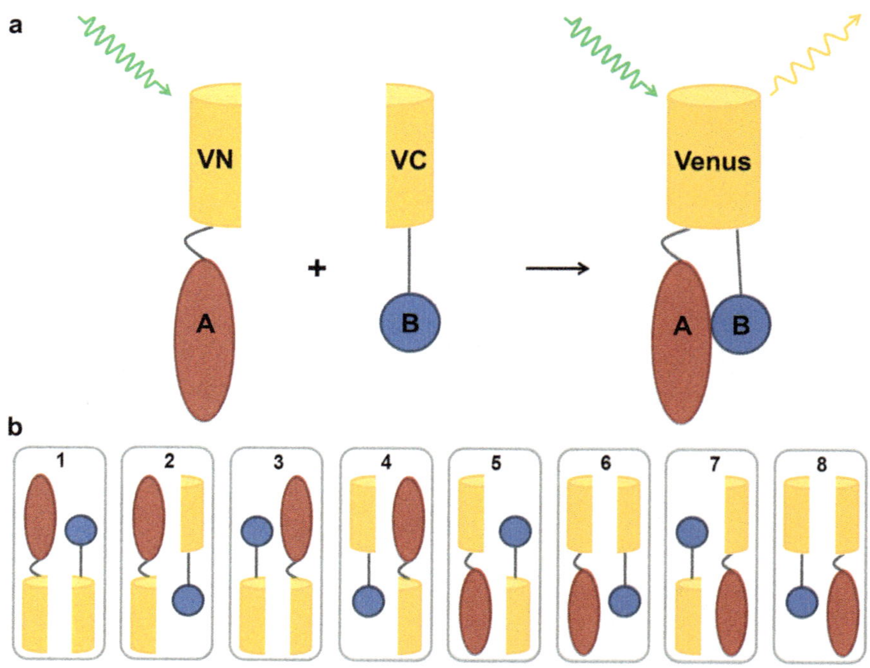

Fig. 1 Principle of BiFC and combinations of fusion proteins. (**a**) Nonfluorescent halves of Venus (Venus N-terminal half (VN) and Venus C-terminal half (VC)) fused to two proteins ("A" and "B") show no emission of yellow light, when excited at 514 nm (*left*). Upon protein interaction, the fluorophore halves are brought in close proximity, the fluorophore reconstitutes, and the excited Venus emits yellow light (max = 528 nm). (**b**) Possible combinations of fusions of VN and VC fused either N (above)- or C-terminally (below) to protein "A" or protein "B." *Gray lines* depict flexible linkers connecting the fusion proteins

2 Materials

2.1 Choice of Vectors and Controls

1. For the analysis of CaM interaction with PSKR1 [3], we chose a Gateway-compatible vector set [4]. Depending on your cloning strategy and the desired application, you may choose between different BiFC vector sets currently available (*see* **Note 1**).
2. Orientation of the fluorophore fusion to the interaction partners: Fig. 1b shows the eight possible combinations for each interaction pair. Depending on the orientation, the BiFC signals may vary significantly, and therefore all possible combinations should be generally tested to reveal the best signal-to-noise ratios. In case of the LRR receptor kinases, which represent type I membrane proteins, the fluorophore halves were only fused to the C-terminus. Thus, combinations 1–4 would apply if the kinase was protein "A." For CaM, there was no difference whether the fluorophore half is fused N- or C-terminally, as shown in previous analyses [5], and therefore only the N-terminal fusion was used.
3. Receptor-VN combined with VC expressed without fusion can be used as a negative control (*see* **Note 2**).
4. The actual BiFC test comprises the comparison of an interaction of CaM (or other potential interacting partners) with the receptor kinase and mutated receptor kinase.

2.2 Agrobacterium-Mediated Transformation of Nicotiana benthamiana Leaf Cells

1. YEB medium: 0.5% (w/v) tryptone, 0.5% (w/v) yeast extract, 0.5% (w/v) sucrose, 50 mM $MgSO_4$, pH (KOH) 7.8, autoclave.
2. Antibiotics stocks: rifampicin (10 mg/ml in DMSO), ampicillin (100 mg/ml in H_2O), kanamycin (100 mg/ml in H_2O). Antibiotics are filtrated, stored in aliquots at −20 °C, and used in 1:1000 dilutions.
3. p19 helper strain [6].
4. Agrobacterium strain (e.g., C58C1).
5. Infiltration buffer: 10 mM MES, 10 mM $MgCl_2$, pH (KOH) 5.7, filtrate, store at 4 °C.
6. Acetosyringone: prepare stock of 100 mM in DMSO, store at −20 °C.
7. 5–6-week-old *N. benthamiana* plants (*see* Fig. 2e).
8. 1 ml plastic syringe.

2.3 Evaluation of Fluorescence

1. Fluorescence stereo microscope.
2. Fluorodecaline.
3. Double-faced adhesive tape.
4. Cork borer (different sizes).

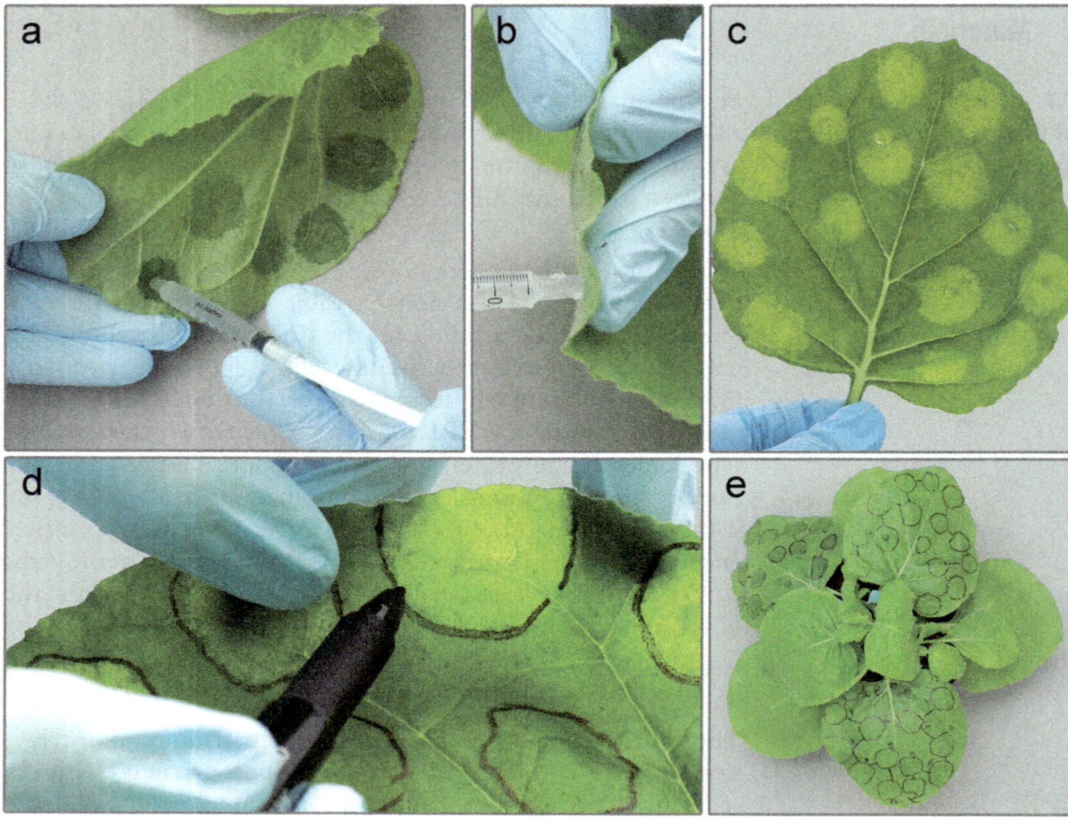

Fig. 2 Infiltration of *N. benthamiana*. (**a**) *Agrobacteria* are infiltrated into the lower side of a *N. benthamiana* leaf. (**b**) Gentle pushing with one finger on top of the leaf against the syringe on leaf bottom helps to create moderate pressure for infiltration. (**c**) Infiltrated areas are wet and transparent. (**d**) Mark infiltrated areas. (**e**) Several inoculated leaves on one plant

5. Confocal laser scanning microscope.
6. 96 well plate, black, flat.
7. Plate reader for excitation/detection of yellow fluorescence.

3 Methods

All procedures are performed at room temperature if not indicated otherwise.

3.1 Infiltration

1. Prepare two cultures of *Agrobacterium*, each harboring one of the two BiFC (VN and VC) plasmids, respectively, in 3 ml YEB medium with appropriate antibiotics. Prepare a flask with appropriate amount of p19 helper strain in YEB with rifampicin, ampicillin, and kanamycin. Incubate overnight at 29 °C on a shaker (*see* **Note 3**).

2. Determine the OD_{600} of each culture.
3. Add the calculated amount of culture in a 2 ml reaction tube that you need to get $OD_{600} = 1$ after adding 1.5 ml of infiltration buffer to the pellet of cells (*see* **Note 4**).
4. Centrifuge at 3200 × *g* for 5 min.
5. Add acetosyringone to a final concentration of 100 μM to the infiltration buffer (calculate the volume you need; *see* **Note 5**).
6. After centrifugation, decant medium from cells and resolve pellet in 1.5 ml infiltration buffer (incl. Acetosyringone).
7. Proceed to **steps 3–6** similar with the p19 strain with the calculated amount you will need in the end (*see* **Note 5**).
8. Mix combinations of VN-harboring agrobacteria with VC-harboring agrobacteria to a total of 3 ml.
9. Add 3 ml of p19 to each combination to a total of 6 ml.
10. Prepare one 6 ml sample with p19 only (to test for background control).
11. Water *N. benthamiana* prior to infiltration to promote opening of stomata.
12. Incubate for 2–3 h.
13. Inoculate tobacco leaves by putting a 1 ml plastic syringe on a plane area on the lower side of the leaf and pressing a finger on the opposite side of the leaf while injecting (*see* Fig. 2a and b; **Note 6**).
14. Get all samples on one leaf to enable comparison within one leaf.
15. Infiltrate several leaves (10–15) to get enough technical replicates.
16. Incubate infiltrated tobacco plants in the greenhouse or growth chamber for 2 days.

3.2 Microscopy

1. Screen all leaves on a stereo microscope for fluorescence and choose one for microscopy (*see* **Note 7**).
2. Put double-faced adhesive tape on an objective slide (*see* Fig. 3e).
3. Punch out small leaf discs of each sample with a size 1 cork borer (*see* **Notes 8** and **9**).
4. Place leaf discs with lower side pointing up (facing to objective) on the tape.
5. Add a droplet of perfluorodecalin on each leaf disc and pierce it with fine tweezers to facilitate the intrusion of the chemical (*see* **Note 10**).
6. After approximately 2 min, add some water and place a cover slip on the sample(s).

Fig. 3 Evaluation of BiFC data. (**a**) Lower leaf site: Infiltrated areas are no longer visible after 2 days. (**b**) Leaf discs punched out with a cork borer. (**c**) Leaf discs are placed with the lower side pointing up (facing the detector) into a 96 well plate. (**d**) Upper leaf site: Syringe contact sites can be detected as circles of dead cells. (**e**) Preparation of samples for microscopy

7. Take pictures with a confocal laser scanning microscope. Settings for Venus fluorescence: excitation with 514 nm laser light, detection window 530–550 nm. 20–40× objectives are sufficient (*see* **Note 11**).

3.3 Quantification of BiFC Fluorescence

1. Add 80 μl of water to each well of a 96 well plate (or several plates). The water will prevent the leaf discs from drying out.
2. Punch out one leaf disc of each sample with a size 3 cork borer (*see* Fig. 3b; **Note 8**). To determine the background fluorescence level, you may use a disc of a p19-infiltrated area.
3. Carefully insert each leaf disc with the lower side pointing up into a well (if the fluorescence is detected from above) with the help of fine tweezers (*see* Fig. 3c; **Notes 12** and **13**).
4. Measure relative fluorescence levels in a plate reader capable of exciting and detecting yellow fluorescence (e.g., a TECAN infinite F200 (TECAN, Suisse)).

5. Analyze the data by calculating mean values and statistics for each leaf and for all samples (*see* **Notes 14**). Apply a *t*-test to determine significance of differences between CaM binding to receptor kinase and receptor kinase with a mutation in the CaM-binding domain. Results can be illustrated in bar charts.

4 Notes

1. Since the reassembly of GFP has been described and plant application was established, the technique of fluorescence complementation was further developed [2, 7, 8]. To ensure similar expression levels of the BiFC partners, they have been assembled onto one expression plasmid. In addition, this plasmid contains a gene coding for a different fluorophore, which is used for BiFC signal normalization [9]. This ratiometric BiFC (rBiFC) allows the comparison of tissues with different expression levels. Further developments include the reduction of background and false-positive signals by splitting the fluorophore at a different position and introducing point mutations, which prevents self-assembly [10].
2. The choice of proper negative controls is important to get reliable results. Horstman and colleagues [11] suggest to use a translational fusion between one fluorophore half and a truncated or mutated version of the interaction protein that is unable to bind its partner as ideal negative control. Indeed, this was the result that was achieved in our studies, comparing the binding ability of the wild-type and mutated PSKR1, providing evidence for the specificity of the BiFC signal. In our studies, we used the soluble VC together with PSKR1-VN as negative control. This combination resulted in very low background signals, which were only marginally higher compared to the p19 control. Caution: The self-assembly can be very strong using only soluble proteins!
3. If you want to test the interaction between the receptor kinase and several CaMs, you might need more and/or larger overnight cultures. Consider that you need the same amount of p19 as the sum of all other cultures together plus 6 ml for background control.
4. The cultures are mostly grown to an OD_{600} over 1. Example with $OD_{600} = 2$: $V1 \times C1 = V2 \times C2$; $(V2 = 1.5\ \text{ml} \times OD_{600} = 1)/OD_{600} = 2 = 0.75$ ml. Add 750 μl *Agrobacterium* suspension to the reaction tube and spin down the cells.
5. You will need 6 ml of infiltration buffer supplemented with acetosyringone per combination.

6. Choose leaves of medium age/size. If you push too hard with the syringe against the opposite finger, you will pierce the leaf; in case you use too low pressure, the bacteria will not enter the leaf but rather spray somewhere else. Safety goggles are therefore strongly recommended. When the bacteria solution enters the leaf, you can observe it distributing in the apoplast (darker circle, Fig. 2a; transparent Fig. 2c and d). Don't bend the leaf too much. Avoid leaf veins, because they are uneven making infiltration difficult, and they will also show more autofluorescence. Dry the leaf surface in between infiltration and at the end with a paper towel. Mark the exact infiltrated area with a marker on the upper side of the leaf (*see* Fig. 2d). Otherwise you will not be able to distinguish the areas 2 days later (*see* Fig. 3a).
7. Depending on the details of your filter set, Venus fluorescence may be observed through the GFP filter.
8. *Nicotiana benthamiana* leaves are not plane even in leaf discs with a diameter of a few mm, which makes microscopy difficult. Prepare 2–3 small sample leaf discs to have some choice of finding a region with most of the cells in focus. As mentioned in **Note 6**, avoid leaf veins due to autofluorescence. Try to get a plane area between the injection site (there might be a circle of dead cells; *see* Fig. 3d) and the outer boundary you tagged with the marker.
9. Prepare only few samples (3) at a time; otherwise, the remaining samples will either dry out or remaining air will rise from the leaf and sum up on the cover slip.
10. Octadecafluorodecahydronaphtalene, also known as perfluorodecalin, displaces air from intercell spaces. You can observe how the leaf gets darker and sometimes air bubbles rising from it.
11. Start with the sample with the strongest fluorescence, to compare the BiFC signal intensities of different samples. Adjust the detector sensitivity adequately avoiding saturation and do not change the detector gain, laser intensity, or magnification for the whole sample series, but only change the 3D axes of the microscope stage to image another sample.
12. Place the leaf disc on the water, not into the water. This ensures them swimming at the same height (and distance to the detector) without reflections or quenching.
13. Put all samples of one leaf together in one row or one column or two adjacent columns.
14. Some leaves don't show any or only very weak fluorescence. That makes the recognition of interactions impossible. Therefore, it is recommended analyzing each leaf separately,

for example, by drawing a graph for each row or column. For best quantitative results, normalization to expression levels is recommended (*see* **Note 15**).

15. Keep in mind that the absence of Venus fluorescence does not necessarily mean that there is no interaction. In this case, expression levels and the localization of the interaction partners should be analyzed. Some vector sets provide the fusion of tags to the BiFC constructs, enabling to test the amount and size of the fusion proteins in a Western blot.

Acknowledgment

We would like to thank Heiner Busch for photography and the Deutsche Forschungsgemeinschaft for financial support.

References

1. Hartmann J, Fischer C, Dietrich P, Sauter M (2014) Kinase activity and calmodulin binding are essential for growth signaling by the phytosulfokine receptor PSKR1. Plant J 78(2): 192–202
2. Hu C-D, Chinenov Y, Kerppola TK (2002) Visualization of interactions among bZIP and Rel family proteins in living cells using bimolecular fluorescence complementation. Mol Cell 9(4):789–798
3. Walter M, Chaban C, Schutze K, Batistic O, Weckermann K, Nake C, Blazevic D, Grefen C, Schumacher K, Oecking C, Harter K, Kudla J (2004) Visualization of protein interactions in living plant cells using bimolecular fluorescence complementation. Plant J 40(3): 428–438
4. Gehl C, Waadt R, Kudla J, Mendel RR, Hansch R (2009) New GATEWAY vectors for high throughput analyses of protein-protein interactions by bimolecular fluorescence complementation. Mol Plant 2(5):1051–1058
5. Fischer C, Kugler A, Hoth S, Dietrich P (2013) An IQ domain mediates the interaction with calmodulin in a plant cyclic nucleotide-gated channel. Plant Cell Physiol 54(4):573–584
6. Voinnet O, Rivas S, Mestre P, Baulcombe D (2003) An enhanced transient expression system in plants based on suppression of gene silencing by the p19 protein of tomato bushy stunt virus. Plant J 33(5):949–956
7. Ghosh I, Hamilton AD, Regan L (2000) Antiparallel leucine zipper-directed protein reassembly: application to the green fluorescent protein. J Am Chem Soc 122(23):5658–5659
8. Kodama Y, Hu CD (2012) Bimolecular fluorescence complementation (BiFC): a 5-year update and future perspectives. Biotechniques 53(5):285–298
9. Grefen C, Blatt MR (2012) A 2in1 cloning system enables ratiometric bimolecular fluorescence complementation (rBiFC). Biotechniques 53(5):311–314
10. Gookin TE, Assmann SM (2014) Significant reduction of BiFC non-specific assembly facilitates in planta assessment of heterotrimeric G-protein interactors. Plant J 80(3):553–567
11. Horstman A, Nougalli Tonaco IA, Boutilier K, Immink RGH (2014) A cautionary note on the use of split-YFP/BiFC in plant protein-protein interaction studies. Int J Mol Sci 15(6):9628–9643

Chapter 15

Pull-down Assay to Characterize Ca^{2+}/Calmodulin Binding to Plant Receptor Kinases

Christine Kaufmann and Margret Sauter

Abstract

Plant receptor-like kinases (RLKs) are regulated by posttranscriptional modification and by interaction with regulatory proteins. A common modification of RLKs is (auto)phosphorylation, and a common regulatory protein is the calcium sensor calmodulin (CaM). We have developed protocols to detect the interaction of an RLK with CaM. The interaction with CaM was shown by bimolecular fluorescence complementation (BiFC) (*see* Chapter 14) and pull-down assay (this chapter). Both methods offer unique advantages. BiFC is useful in showing interaction of soluble as well as of membrane-bound proteins in planta. Pull-down assays are restricted to soluble proteins and provide in vitro data. The pull-down assay provides the advantage that proteins can be modified prior to binding and that experimental conditions such as the concentration of Ca^{2+} or other divalent cations can be controlled. This chapter provides a pull-down protocol to study RLK-CaM interaction with optional steps to investigate the impact of RLK phosphorylation or of Ca^{2+}.

Key words Protein tagging, Calmodulin expression, Calcium, Peptide receptor kinase expression, Protein binding, Phosphorylation, Pulldown

1 Introduction

Plants possess a large number of receptor-like kinases (RLKs) with more than 600 found in *Arabidopsis thaliana* [1]. The brassinosteroid receptor brassinosteroid insensitive 1 (BRI1), the immune receptor flagellin sensing 2 (FLS2), and peptide receptors such as the phytosulfokine receptors 1 and 2 (PSKR1 and PSKR2) belong to the leucine-rich repeat (LRR) RLKs with more than 200 members in *Arabidopsis* that have an extracellular ligand-binding LRR domain [2]. The LRR is linked to an intracellular kinase by a single transmembrane helix. The intracellular receptor part of LRR RLKs can be regulated at various levels including posttranslational modification in the kinase domain, but also in the juxtamembrane region and at the C-terminus, and by binding to regulatory proteins.

Reidunn Birgitta Aalen (ed.), *Plant Receptor Kinases: Methods and Protocols*, Methods in Molecular Biology, vol. 1621, DOI 10.1007/978-1-4939-7063-6_15, © Springer Science+Business Media LLC 2017

For most RLKs, the regulatory mechanisms that control their activity are however not or only partly understood.

Regulation by calcium was suggested for members of subclass X of the LRR RLK family that share a conserved amphipathic α-helix (helix E) in kinase subdomain VIa that is predicted to bind the calcium sensor calmodulin (CaM) [3]. BRI1 and PSKRs are subgroup X members [2]. For BRI1 and PSKR1, binding to CaM in vivo was demonstrated previously by the bimolecular fluorescence complementation (BiFC) method [4]. Further information and a detailed protocol of BiFC can be found in Chapter 15. The methods described here and in Chapter 15 should be useful to clarify if an RLK has the ability to interact with CaM and is hence likely to be regulated by calcium.

As a complementary approach to BiFC to study the RLK-CaM interaction, we developed a pull-down assay using heterologously expressed tagged CaM2 and tagged PSKR1 kinase (Fig. 1). The in vitro system allows for controlled reaction conditions such as the concentration of divalent cations. It is further possible to posttranscriptionally modify the receptor kinase prior to the interaction assay. This approach allowed us to study the impact of Ca^{2+}, of Mg^{2+}, and of kinase phosphorylation on PSKR1-CaM2 binding. An important point to consider is that this experimental setup is not suited for membrane proteins. With regard to RLKs, it is however possible to express and use the cytoplasmic receptor part (PSKR1-KD) that harbors the CaM-binding site as a soluble protein.

2 Materials for Pull-down Assay

2.1 Choice of Vectors

1. It is important that the tag does not alter protein activity. This should be considered before choosing the tags and the position of the tag. If possible, it is best to compare the activity of the tagged protein to that of the untagged protein. For PSKR1, the N-terminal 6×His(H_6) tag does not disturb kinase activity in vitro [5]. N-terminally H_6-tagged fusion protein can be produced using the pETDuet vector system (*see* **Note 1**).
2. Calmodulin (CaM) is a regulatory protein of small size. A FLAG tag at the C-terminus can be used for detection of CaM2. Large quantities of soluble CaM2 can be obtained using the vector system pMAL™ Protein Fusion and Purification System (New England Biolabs). Fusion to the maltose-binding protein (MaBP) enhances solubility and results in higher yields [6].

2.2 Materials Used to Express MaBP-CaM-FLAG

1. *Escherichia coli* strain BL21(DE3).
2. Expression medium: LB medium with 0.2% (w/v) glucose, autoclave, and add ampicillin prior to use to a final concentration of 100 µg/mL. The presence of glucose in the medium is necessary to repress expression of amylase (*see* **Note 2**).

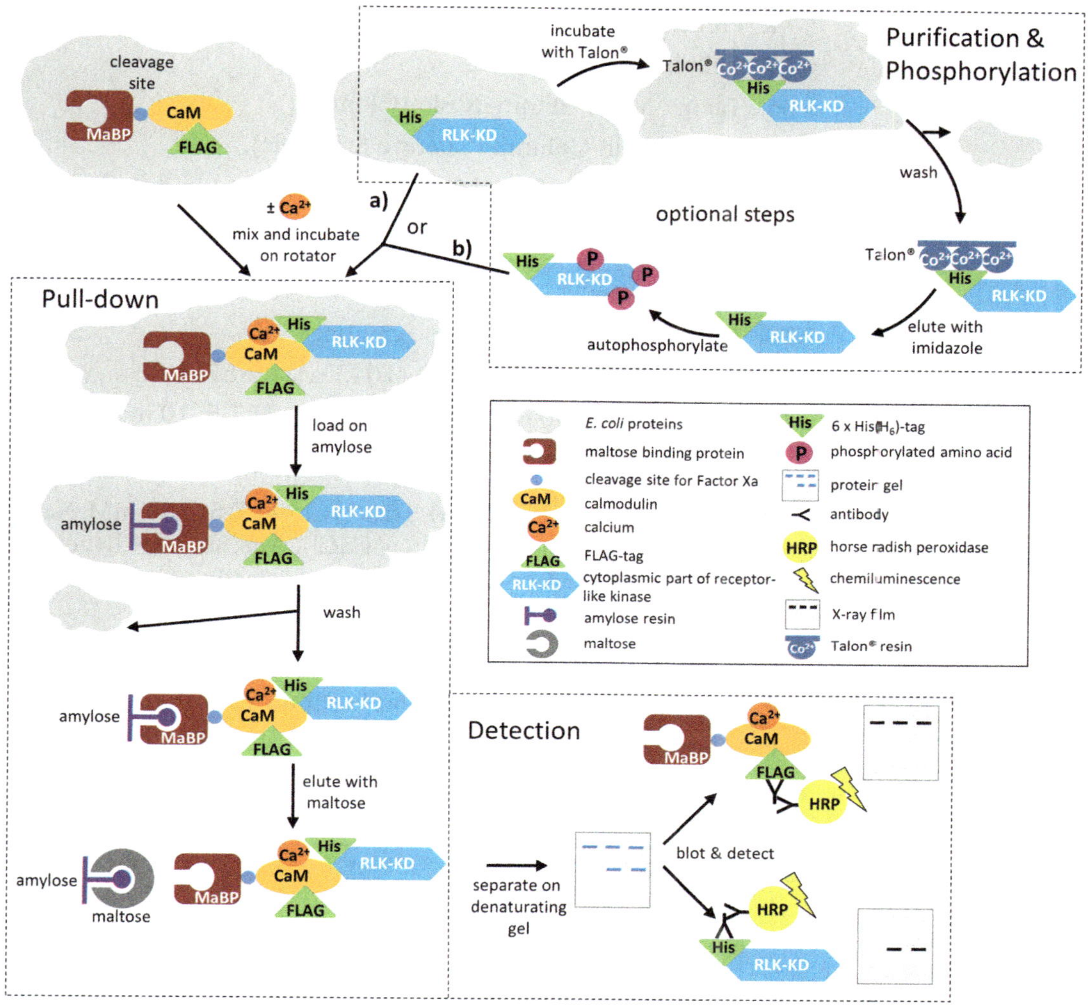

Fig. 1 Schematic overview of the pull-down assay with CaM and RLK-KD fusion proteins. (**a**) Crude *E.coli* extracts with heterologously expressed MaBP-CaM-FLAG and H_6-RLK-KD can be used directly. (**b**) To study the impact of autophosphorylation on CaM binding, H_6-RLK-KD can be affinity purified by Talon® using the 6×His(H_6) tag, followed by incubation with kinase buffer to allow for autophosphorylation. The purified and modified kinase is subsequently incubated with crude extract of MaBP-CaM-FLAG for binding

3. Extraction buffer: 20 mM Tris/HCl pH 7.5, 200 mM NaCl.
4. French press.
5. Cell disruptor.

2.3 Materials for the Expression of the His-Tagged Kinase Domain of a Receptor-like Kinase (H_6-RLK-KD)

1. *Escherichia coli* strain BL21(DE3).
2. Expression medium: LB medium, autoclave; add ampicillin to a final concentration of 100 μg/mL just before use.
3. Extraction buffer: 50 mM sodium phosphate pH 7.0, 300 mM NaCl.

4. French press.
5. Cell disruptor.
6. Talon® Metal Affinity Resin (Clontech).
7. Pierce® Spin Column (Thermo Scientific).
8. Washing buffer: 50 mM sodium phosphate pH 8.0, 300 mM NaCl, 5 mM imidazole.
9. Elution buffer: 50 mM sodium phosphate pH 8.0, 300 mM NaCl, 150 mM imidazole.
10. NanoDrop 2000 (Thermo Scientific).
11. Roti-Spin Mini-10 MWCO (10 kDa) (Carl Roth).
12. Kinase buffer: 50 mM HEPES/KOH pH 7.5, 10 mM $MnCl_2$, 1 mM dithiothreitol, 0.2 mM ATP.

2.4 Materials for Pulldown

1. Interaction buffer: 20 mM Tris/HCl pH 7.5, 200 mM NaCl. Supplement with ions (Ca^{2+} as $CaCl_2$, Mg^{2+} as $MgCl_2$), chelators (e.g., EDTA or EGTA), or other cofactors as needed.
2. Rotator.
3. Amylose resin (NEB).
4. Pierce® Spin Column (Thermo Scientific).
5. Elution buffer: 20 mM Tris/HCl pH 7.5, 200 mM NaCl, 10 mM maltose.

3 Pull-down Methods

3.1 Expression of MaBP-CaM-FLAG

1. Inoculate a flask with 30 mL of expression medium with cells of a verified MaBP-CaM-FLAG containing clone and incubate on a shaker at 37 °C.
2. Determine the OD_{600} of the culture and proceed when culture reaches OD of 0.5.
3. Induce protein expression by adding isopropyl β-D-1-thiogalactopyranoside (IPTG) to a final concentration of 0.3 mM to the culture and shake at 37 °C for 2 h.
4. Harvest the cells by centrifugation at 4000 × *g* for 5 min at 4 °C. Discard the supernatant and resolve the pelleted cells in 4 mL extraction buffer.

3.2 Expression of H_6-RLK-KD

1. Inoculate a flask with 50 mL of expression medium with cells of a verified H_6-RLK-KD containing clone and shake at 37 °C.
2. Determine the OD_{600} of each culture. At an OD_{600} of 0.5 let bacteria cool down to room temperature to avoid formation of inclusion bodies (*see* **Note 3**).

3. Induce protein expression by adding IPTG to a final concentration of 1 mM to the culture and shake at room temperature for 16 h.
4. Harvest the cells by centrifugation at 4000 × *g* for 5 min at 4 °C. Discard the supernatant and resolve the pelleted cells in 3.5 mL extraction buffer.

3.3 Isolation of MaBP-CaM-FLAG and H_6-RLK-KD

1. Disrupt the resuspended cells in a French press with 20,000 PSIG cell pressure in a precooled cell.
2. Sonicate the solution three times for 10 s with appropriate energy amplitude setting for used microtip to destroy accumulated DNA, while having the sample kept cooled in an ice bath (*see* **Note 4**).
3. Centrifuge the mixture for 30 min at 21,000 × *g* and 4 °C.
4. Collect the supernatant, discard the pellet, and freeze MaBP-CaM-FLAG at −80 °C until use or use directly for pulldown. Instead of using crude extract, MaBP-CaM-FLAG can be purified and modified prior to pulldown (*see* **Note 5**). For H_6-RLK-KD, use crude extract immediately or proceed with affinity purification (*see* Subheading 3.4).

3.4 Purification (See Note 6) and Autophosphorylation (See Note 7) of H_6-RLK-KD (Optional Steps)

1. Mix 50 μL of Talon resin with 500 μL of extraction buffer in a Pierce® Spin Column and centrifuge for 30 s at 400 × *g* in a tabletop centrifuge at room temperature to wash out resin storage solution. Discard the flow-through.
2. Incubate the washed Talon resin with the prepared crude extract for 1 h at room temperature in a rotator.
3. Wait for a few minutes until the resin sediments to the bottom of the container by gravity or centrifuge for 1 min at 400 × *g* at room temperature.
4. Discard the supernatant except for 500 μL. Mix the pelleted resin with the leftover 500 μL solution and load onto the Spin Column.
5. Centrifuge for 30 s at 400 × *g* in a tabletop centrifuge at room temperature. Discard the flow-through.
6. Wash three times with 500 μL of washing buffer.
7. Elute the bound H_6-RLK-KD by adding 50 μL of elution buffer and incubate for 10 min prior to centrifugation.
8. For ultrafiltration, transfer the eluted H_6-RLK-KD to a Roti®-Spin Column (*see* **Note 8**), centrifuge at 12,000 × *g* at 4 °C in a tabletop centrifuge to reduce sample volume, and discard the flow-through. Add 10× the sample volume of 50 mM HEPES/KOH pH 7.5 to the Roti®-Spin Column to adjust to the kinase buffer, centrifuge, and discard flow-through. Repeat

this procedure two to three times. Centrifuge to reduce the sample to half the volume desired. Pipette the solution from the membrane to a fresh tube. Wash residual protein from the membrane with the same volume of buffer and combine.

Determine the protein concentration with a NanoDrop by measuring the absorbance at 280 nm.

9. For kinase autophosphorylation, incubate H_6-RLK-KD for 1 h at 25 °C in kinase buffer.
10. Before using autophosphorylated kinase for pulldown, the buffer conditions need to be adjusted. To do so, ultrafiltrate and concentrate the autophosphorylated kinase protein in a Roti®-Spin Column as described above except that here the kinase buffer is replaced by interaction buffer.

3.5 Pulldown

1. Estimate the protein concentration of the proteins to be analyzed in the respective crude extracts by running an aliquot on an SDS-PAGE together with a standard, e.g., BSA, of known protein concentration (*see* **Note 9**).
2. Incubate equivalent amounts of H_6-RLK-KD and MaBP-CaM-FLAG crude extracts in incubation buffer in a total volume of 250 μL. We used ~50 μg protein each in our assays. Alternatively, use the purified and autophosphorylated H_6-RLK-KD. Binding of CaM may be influenced by calcium ions. To test that, add 100 μM $CaCl_2$ to the buffer or add 100 μM $MgCl_2$ as a control.
3. Place the sample(s) at room temperature for 30 min on a rotator to allow for binding.
4. Mix 25 μL of amylose resin with 300 μL of interaction buffer in a Spin Column and centrifuge for 1 min at 400 × *g* at room temperature in a tabletop centrifuge to wash out resin storage solution. Discard the flow-through. Repeat this washing procedure two times.
5. Add the washed amylose resin to the sample. Incubate for another 30 min on a rotator to allow for binding of the complex via MaBP fusion protein to amylose resin.
6. Centrifuge the mixture in a Spin Column at 400 × *g* for 1 min at room temperature. Collect the flow-through and store at 4 °C for subsequent analysis on a gel.
7. Wash three times with 300 μL interaction buffer ± ions as described above.
8. For elution, mix the amylose resin with 2× the volume of amylose resin used initially (in this case 50 μL), incubate for 5 min, and centrifuge at 400 × *g* for 1 min at room temperature.
9. Separate the protein complex on a denaturating protein gel, e.g., 12.5% SDS-gel.

10. Blot the proteins on a polyvinylidene difluoride (PVDF) membrane and detect with appropriate primary and secondary antibodies, e.g., anti-FLAG and anti-His primary antibodies and horse radish peroxidase (HRP)-conjugated anti-mouse secondary antibody. Detect proteins by exposing the membrane to an X-ray or chemiluminescence film to visualize chemiluminescence. Protein detection by Western blot analysis should be optimized prior to the pull-down experiment using crude protein extract (*see* **Note 10**).

4 Notes

1. In general, the Duet Vectors from Novagen have advantages if protein complexes are to be studied by coexpression. For the pull-down assay described here, any kind of vector system can be used that works properly for protein expression and purification of H_6-tagged fusion proteins.
2. Repression of amylase expression is important. Amylase degrades the amylose resin during purification and thereby leads to a lower yield of the desired fusion protein.
3. Heterologously expressed proteins can form insoluble, inaccessible and nonfunctional aggregates termed inclusion bodies. There are several ways to avoid inclusion body formation. In general, it all comes down to slow down the speed at which fusion proteins are made [7]. This can be achieved by using a low-copy number vector, by coexpression of molecular chaperones, induction with a lower IPTG concentrations (<1 mM), use of alternative media that are rich (e.g., 2YT) or less rich (e.g., LB), the duration of bacterial growth after induction, and—as described in this protocol—by lowering the temperature at which the bacteria are grown before and after induction [8].
4. Accumulated DNA increases the viscosity of the solution and blocks membranes and therefore impairs purification or interactions of proteins. Sonication of the solution leads to an increased temperature which may damage the proteins in the sample. Therefore, cooling the sample while sonicating is highly recommended. Additionally, perform sonication in three intervals to allow the mixture to cool down in between and choose a sonication energy amplitude setting as low as possible to avoid destruction of proteins.
5. As an optional step, it is possible to purify MaBP-CaM-FLAG with amylose resin followed by elution with maltose. This may be helpful if no interaction is observed after pulldown with crude extract. This result can be caused by interference by

MaBP. To exclude this scenario, MaBP can be cleaved off with Factor Xa resulting in a CaM-FLAG fusion protein. Cross-link anti-FLAG antibodies to Protein A beads, e.g., with the Pierce® Crosslink Immunoprecipitation Kit and proceed with a co-immunoprecipitation (Co-IP). The cross-link prevents release of antibodies when proteins are eluted. The heavy and light antibody chains may interfere with immunoblotting and immunodetection of the proteins of interest. In this case, a Co-IP with cross-linked anti-His antibodies can be used to confirm a specific CaM-RLK interaction.

6. If no interaction is observed, this may be due to interfering proteins in the *E. coli* crude extract. In this case, one or both interaction partners can be purified from their respective crude extracts. Pulldown with crude extracts however is preferred, because it supports the specificity of the interaction.
7. Posttranslational modifications, e.g., phosphorylation can alter the protein structure. In some cases, a conformational change is required for an interaction or may enhance or decrease the interaction. Therefore, when working with RLKs, this should be kept in mind and considered for analysis. To examine the effect of autophosphorylation of an RLK, the RLK is purified via its tag and subsequently incubated in kinase buffer to allow for autophosphorylation.
8. The Roti®-Spin Column is a commonly used centrifugal filter. You can also use alternative size exclusion filters, e.g., Centricon from Millipore. The important thing to keep in mind is to choose the appropriate size exclusion limit. This is dependent on the size of the protein to be analyzed. The sample is placed in the reservoir above the membrane. During centrifugation, proteins smaller than the specified cutoff pass the membrane pores while the protein of interest is retained in the reservoir. With a filter or membrane column, you can easily concentrate a protein sample or switch buffer conditions.
9. When working with crude extract, quantification of the expressed target protein with Bradford or other photometric methods is not possible. To determine the amount of target protein(s), an aliquot can be run on an SDS gel. To estimate protein concentration, a dilution series of a known protein, e.g., bovine serum albumin (BSA), can be included on the gel.
10. To make sure that the interacting proteins are easily detected by Western blot, optimize the 1° and 2° antibody concentrations. Begin by using the concentration suggested by the manufacturer and use higher (if no signal is detected) or lower (when the background is high) concentrations in steps of twofold. As primary antibodies, we used a monoclonal anti-FLAG M2 (Sigma) antibody diluted 1:2500 or a 6×–His epitope tag

antibody His.H8 also diluted 1:2500 (Thermo Fisher). As secondary antibodies, horse radish peroxidase-conjugated anti-mouse antibodies diluted 1:50,000 (Life Technologies) was used. To visualize the proteins, we used the ECL Plus Western blotting substrate (Pierce).

Acknowledgment

We would like to thank the Deutsche Forschungsgemeinschaft for financial support.

References

1. Shiu S, Bleecker A (2001) Receptor-like kinases from Arabidopsis form a monophyletic gene family related to animal receptor kinases. Proc Natl Acad Sci U S A 19:10763–10768
2. Diévart A, Clark S (2003) Using mutant alleles to determine the structure and function of leucine-rich repeat receptor-like kinases. Curr Opin Plant Biol 6(5):507–516
3. Oh M, Kim H, Wu X, Clouse S, Zielinski R, Huber S (2012) BRASSINOSTEROID-INSENSITIVE 1 receptor kinase provides a possible link between calcium and brassinosteroid signalling. Biochem J 443(2):515–523
4. Hu C-D, Chinenov Y, Kerppola TK (2002) Visualization of interactions among bzip and rel family proteins in living cells using bimolecular fluorescence complementation. Mol Cell 9(4):789–798
5. Hartmann J, Fischer C, Dietrich P, Sauter M (2014) Kinase activity and calmodulin binding are essential for growth signaling by the phytosulfokine receptor PSKR1. Plant J 78(2):192–202
6. New England Biolabs (2005) pMAL™ protein fusion & purification system, instruction manual. https://www.neb.com/~/media/Catalog/All-Products/FF91B67397D945E0956F3483174CF7E3/Datacards%20or%20Manuals/Manual%20E8200.pdf
7. Fakruddin M, Mazumdar RM, Bin Mannan KS, Chowdhury A, Hossain MN (2013) Critical factors affecting the success of cloning, expression, and mass production of enzymes by recombinant *E. coli*. ISRN Biotechnol 2013:590587. doi:10.5402/2013/590587
8. Sanchez de Groot N, Ventura S (2006) Effect of temperature on protein quality in bacterial inclusion bodies. FEBS Lett 580:6471–6476

Part V

Imaging of PRK Dynamics

Chapter 16

The Use of FRET/FLIM to Study Proteins Interacting with Plant Receptor Kinases

Stefanie Weidtkamp-Peters and Yvonne Stahl

Abstract

The investigation of protein interactions in living plant tissue has become of increasing importance in recent years. A high spatial and temporal resolution for the observation of *in vivo* protein interaction is needed, e.g., in order to follow changes of plant receptor kinase interactions and complex formation over time. *In vivo* fluorescence or Förster resonance energy transfer (FRET) measurements allow for detailed analyses of interacting proteins in their natural environment at a subcellular level. Especially FRET-FLIM (fluorescence lifetime imaging microscopy) measurements provide an extremely powerful and reliable tool meeting the demands for investigating *in vivo* protein interaction quantitatively and with high precision. Here, we will describe in detail how to practically perform *in vivo* FRET measurements of receptor kinases in plants and discuss potential pitfalls and points of consideration.

Key words FRET, APB, FLIM, Fluorescent proteins, Protein interaction

1 Introduction

The use of fluorescent protein (FP)-tagged proteins in plants has become an invaluable tool in cell biology to observe protein localization, dynamics, and function. Despite the relatively high autofluorescence levels in plants compared to other organisms, the increasing sensitivity of available imaging systems allows for the detection of even lowly abundant FP-labeled proteins within their native surroundings. Live cell imaging approaches have been shown to extend the knowledge on proteins gained by biochemical and genetic studies on tissue, cellular, and subcellular level. However, proteins rarely act on their own, so the ability to measure protein-protein interaction *in vivo* has become of increasing importance. The existence of different FP variants and their co-expression in the same cells has led to the possibility to study protein complexes and their dynamic behavior *in vivo*. Co-localization experiments of differently FP-labeled proteins can give indications on the probability of protein complexes. Nevertheless, subcellular

Reidunn Birgitta Aalen (ed.), *Plant Receptor Kinases: Methods and Protocols*, Methods in Molecular Biology, vol. 1621, DOI 10.1007/978-1-4939-7063-6_16, © Springer Science+Business Media LLC 2017

co-localization is only a prerequisite for protein-protein interaction as it is restricted by the diffraction limit (~250 nm) of a (confocal) microscopic setup and therefore does not necessarily indicate physical interaction of the co-localizing proteins. Fluorescence or Förster resonance energy transfer (FRET) is a method to measure the interaction of proteins that are in close proximity (<10 nm) with high precision beyond the diffraction barrier.

1.1 FRET

Interaction of proteins in living cells and tissues can be investigated by using FRET. In case of FRET, energy is transferred from an excited donor fluorophore to an acceptor fluorophore. The phenomenon of FRET was first described by Theodor Förster in 1948 [1]. Prerequisites for the energy transfer are an overlap of the emission spectrum of the donor fluorophore and the excitation spectrum of the acceptor fluorophore (Fig. 1), the very close vicinity of the two fluorophores (<10 nm), as well as the correct orientation of the dipole moments of both fluorophores. The energy transfer leads to measurable changes in fluorescence parameters of the donor and acceptor fluorophore: fluorescence intensity and fluorescence lifetime of the donor fluorophore decrease in the presence of FRET, and fluorescence intensity of the acceptor fluorophore increases.

The distance between two fluorophores leading to an energy transfer with the efficiency of 50% from the donor to the acceptor is called Förster radius (R_0). The Förster radius of the FRET pair EGFP (donor) and mCherry (acceptor) is 5.1 nm [2]. The FRET transfer rate (k_{FRET}) can be described as a function of the distance between donor and acceptor (R_{DA}) and depends on the donor-to-acceptor separation distance with an inverse sixth power law due to the dipole-dipole coupling mechanism [3].

$$k_{\mathrm{FRET}} = \frac{1}{\tau_{\mathrm{D}(0)}}\left(\frac{R_0}{R_{\mathrm{DA}}}\right)^6 \tag{1}$$

Due to this strong dependence on the separation distance, small changes in distance, e.g., due to conformational changes of a protein, are able to cause large changes in FRET, which can be detected by mainly three different microscopic methods:

1. Sensitized emission of acceptor fluorescence.
2. Acceptor photobleaching (APB).
3. Fluorescence lifetime imaging microscopy (FLIM).

Sensitized emission and acceptor photobleaching can be carried out using conventional widefield or confocal microscopes, while FLIM measurements can only be carried out at a microscope that is additionally equipped with a pulsed laser source, single photon-sensitive detectors, and photon counting electronics. Sensitized

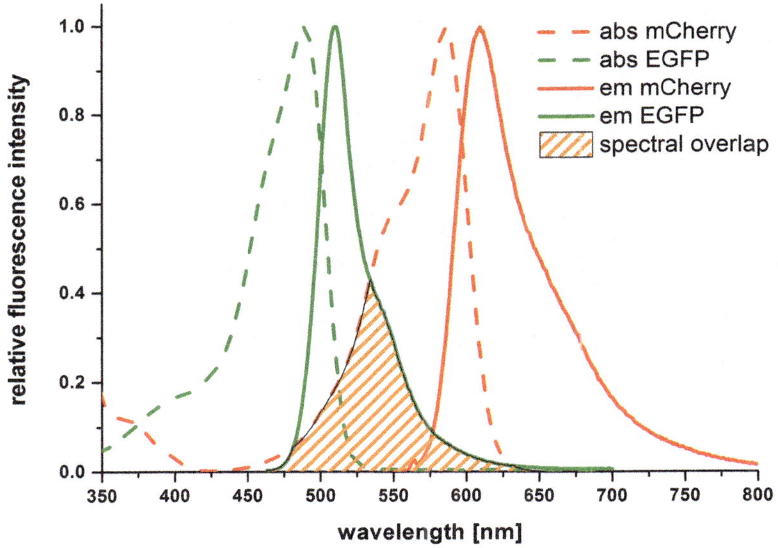

Fig. 1 Absorption (abs) and emission (em) spectra and spectral overlap (orange) of the FRET pair EGFP/mCherry

emission measurements can cause difficulties in plants due to, e.g., spectral cross talk and autofluorescence. FRET-APB is a good way to measure FRET, but requires the bleaching of the acceptor fluorophore, and therefore no repetitive measurements can be done on the same cell. Measuring FLIM has the advantage of using only low excitation power thereby not inducing bleaching artifacts. Furthermore, FLIM can be used to look at the dynamics of complexes over time with a high spatial and temporal resolution.

1.2 FRET Measurements Using Acceptor Photobleaching (FRET-APB)

In case of FRET, the fluorescence intensity of the donor fluorophore is quenched by the acceptor fluorophore. If the acceptor fluorophore is irreversibly destroyed by a strong laser pulse (bleaching), the energy transfer from donor to acceptor is blocked. Thus, the fluorescence intensity of the donor fluorophore will increase. The advantage of this straightforward method to measure FRET between two fluorophores is the direct fluorescence intensity-based readout (Fig. 2).

FRET-APB measurements can be easily quantified by measuring the intensity increase of the donor fluorophore after bleaching of the acceptor according to the following formula:

$$FRET\ efficiency\ E = \frac{Intensity_{Donor\ after} - Intensity_{Donor\ before}}{Intensity_{Donor\ after}} * 100 \quad (2)$$

However, this formula does not include any corrections for background fluorescence or effects like fluorescence recovery of the donor fluorophore during bleaching time of the acceptor.

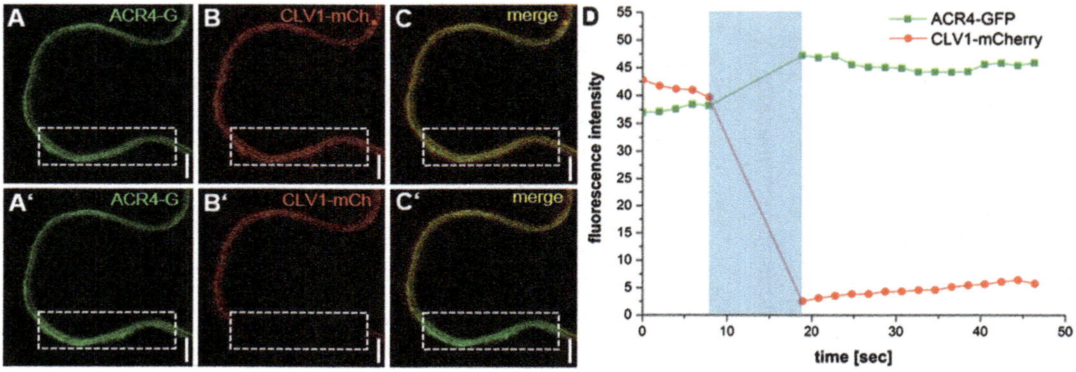

Fig. 2 FRET-APB experiment in transiently expressing *N. benthamiana* leaf epidermis. ACR4-GFP and CLV1-mCherry intensity before (**a**–**c**) and after bleaching (**a**′–**c**′) of the acceptor in the *white rectangle*. (**d**) The increase of the GFP (donor) intensity after bleaching mCherry (acceptor) in the *blue-shaded area* is clearly visible. Scale bars represent 5 μm

Therefore, a set of control experiments is necessary to validate the FRET efficiency *E*. These control experiments include the performance of the same bleaching experiments only with the donor fluorophore present. This will reveal the recovery rate of the donor during the bleaching step [2, 4].

1.3 FRET Measurements Using Fluorescence Lifetime Imaging Microscopy (FRET-FLIM)

Fluorescence lifetime imaging microscopy is based on the measurement of arrival times of single photons at a detector after their excitation with a pulsed laser (Fig. 3a). This method of time-correlated single photon counting (TCSPC) requires, in addition to a pulsed laser source, single photon-sensitive detectors like avalanche photodiodes (APDs) or hybrid detectors (HyDs) and photon counting electronics. The counting board acts like a precise stopwatch and registers the arrival time of single photons over multiple cycles.

The individual photon counts are summed up in time channels resulting in a fluorescence lifetime histogram with exponential decay behavior (Fig. 3b). The mean fluorescence lifetime calculated from this decay histogram is one characteristic parameter of a certain fluorophore that depends on its photophysical properties. Changes in fluorescence lifetime can be monitored with high sensitivity via TCSPC. These changes report on changes in the environment of the fluorophore, due to FRET, but also other influences, e.g., a decrease in pH leads to increased protonation of some fluorescent proteins resulting in a dark state of the fluorophores and a decrease in fluorescence lifetime [5].

In case of FRET, the mean fluorescence lifetime is decreased due to the fact that the resonance energy transfer occurring between two neighboring fluorophores is another way for the excited donor fluorophores to dispose of their excitation energy. This results in a faster depopulation of the excited state and

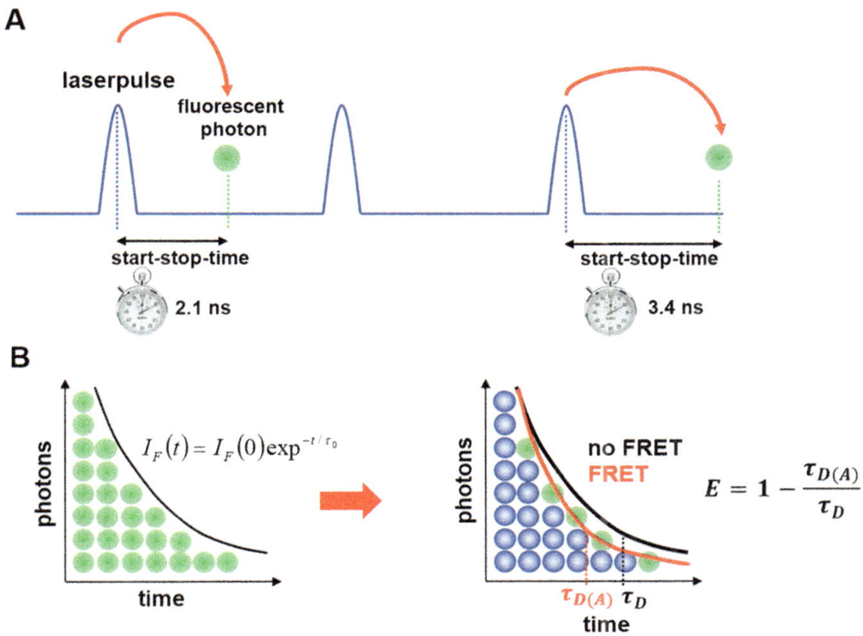

Fig. 3 Fluorescence lifetime imaging. (**a**) Time-correlated single photon counting (TCSPC) using pulsed laser excitation and start-stop electronics. (**b**) Fluorescence lifetime histogram in case of no FRET (*black*) and FRET (*red*). Modified from [5]

therefore shortening of the mean fluorescence lifetime of these fluorophores (Fig. 3b).

In contrast to intensity-based measurements like sensitized emission and acceptor photobleaching described earlier for FRET, FRET-FLIM measurements allow for a more detailed analysis as the fitting of the individual decay curves of fluorescence lifetime measurements can, e.g., identify the amount of interacting and noninteracting molecules [6]. If enough photon counts are collected in single pixels, this analysis can be carried out with very high temporal and spatial resolution, giving additional insights in, e.g., differential receptor kinase interaction at different subcellular microdomains [7–9]. An example of a donor-only and FRET sample is shown in Fig. 4. Here, the high spatial resolution of FRET-FLIM measurements allows detecting FRET at differential subcellular microdomains such as plasma membrane and plasmodesmata.

1.4 Which FRET Pair to Choose?

ECFP and EYFP is the most common FRET pair in use due to their high spectral overlap. However, ECFP as a donor fluorophore is in many cases very sensitive to photobleaching and is less bright than, e.g., EGFP due to a low quantum yield [10]. In FLIM measurements the multi-exponential decay of ECFP complicates FRET-FLIM data analysis. An alternative FRET pair in same

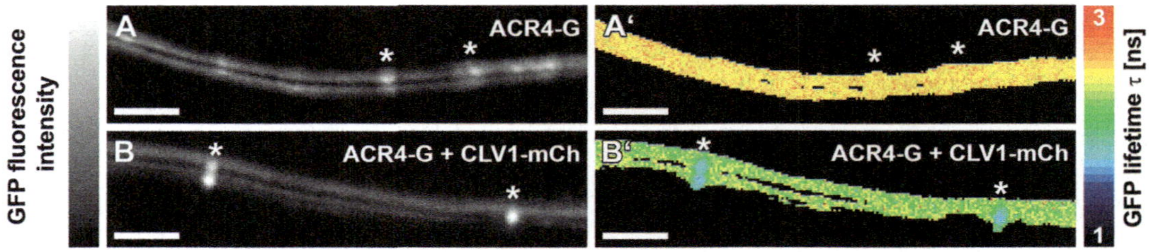

Fig. 4 GFP fluorescence intensity and lifetime. (**a**–**b**′) Fluorescence intensity images of ACR4-GFP (**a**) and ACR4-GFP + CLV1-mCherry (**b**) localizing at the plasma membrane and plasmodesmata of transiently expressing *N. benthamiana* leaf epidermal cells. (**a**′, **b**′) Fluorescence lifetime images of the same cells shown in (**a**) and (**b**). GFP fluorescence intensities and GFP fluorescence lifetime τ [ns] are color-coded. *Asterisks* mark the localization of plasmodesmata. Scale bars represent 2 μm. Modified from [7]

spectral range but with higher brightness is mCerulean in combination with mVenus [11]. For measuring FRET of plasma membrane-localized receptor kinases, yet a different FRET pair is favorable, as the blue autofluorescence of the nearby cell wall will complicate intensity and lifetime-based FRET measurements in plants.

GFP variants like EGFP, EYFP, and mVenus in combination with red fluorescent proteins, e.g., mRFP and mCherry, are good alternatives and more reliable regarding their photophysical properties. In general, one has to consider the autofluorescent background depending on the tissue or even subcellular compartment of plants when choosing an appropriate FRET pair.

Furthermore, all FP-fused proteins have to be tested for correct localization and ideally for functionality beforehand.

2 Materials

2.1 Equipment and Solutions

1. Confocal microscope with continuous wave lasers (for FRET-APB) or confocal microscope equipped with pulsed lasers, single photon sensitive detectors, and TCSPC (time-correlated single photon counting) electronics (for FLIM).
2. Power meter to measure the power of the excitation light at the objective.
3. Erythrosine B solution (*see* **Note 1**).
4. Fluorescence-free water (e.g., Milli-Q).
5. Immersion water or other immersion liquid (depending on objective used).
6. FP-expressing plant material.
7. Razor blades, microscope slides, coverslips (#1.5, 170 μm thickness), and forceps.

8. Black box to cover sample for FLIM measurements.

2.2 Plant Material

The plant material used for FRET-APB and FRET-FLIM measurements is the same. For one set of measurements, the following plant material is needed:

1. "Donor-only" sample is required as everyday reference (e.g., ACR4-GFP).
2. FRET sample expressing both donor- and acceptor-labeled proteins (e.g., ACR4-GFP + CLV1-mCherry).
3. A positive control consisting of a fusion of donor and acceptor (e.g., ACR4-GFP-mCherry). This positive control is mandatory for FRET-APB measurements to adjust the acquisition and bleach settings for the green and red channel.
4. If you do not have stably expressing plant lines, you can use transient expression in *Nicotiana benthamiana*. Here, an inducible expression of the FP-labeled receptor kinases, e.g., by estradiol induction, is advisable, because of possible overexpression artifacts when using an expression system employing a constitutive promoter instead. The time needed for expression has to be determined empirically for every used FP-labeled receptor kinase [12].

3 Methods

3.1 Mounting of Plant Material

Prepare your plant tissue (e.g., root or leaf epidermis) in water (in this case using a water immersion objective is best). When preparing leaf epidermis, cut out a square (0.5 × 0.5 cm) of the leaf and place the abaxial site of the leaf facing the coverslip. To get rid of the air trapped in the leaf tissue, carefully use a razor blade to apply moderate pressure to the leaf by moving over the coverslip until the leaf becomes slightly transparent (Fig. 5). Alternatively, you can infiltrate water into the leaf to get rid of the air, but still you will have to carefully flatten the leaf as described above. Take care not to destroy cells, as this will result in increased autofluorescence.

3.2 Calibration Measurements for FRET-FLIM

Calibration measurements should be done for every measurement day to monitor the performance of your FLIM system. Generally, it is important to prevent the detection of any room light by putting a black box over your samples.

1. Set your pulsed excitation laser to an appropriate repetition rate (e.g., 50, 40, or 32 MHz). For measuring EGFP, YFP, or mVenus, we recommend using a repetition rate of 32 MHz in order to record the full fluorescence lifetime decay of the donor FP. If possible, measure the instrument response function (IRF) (*see* **Note 2**) of your system using a 10 s point

Fig. 5 *Nicotiana benthamiana* leaf preparation. (**a**) Typical growth state of *N. benthamiana* leaves used for transient expression of FP-labeled proteins. (**b**) ca. 0.5 cm × 0.5 cm leaf pieces on microscope slide with coverslip mounted in water before (*right*) and after (*left*) carefully removing air out of the leaf by applying pressure. Scale bars represent 2 cm in (**a**) and 1 cm in (**b**)

measurement with erythrosine B (*see* **Note 1**) or the reflection of the laser with a mirror and appropriate optical density filters in front of your detectors instead of emission filters. If you are not able to record an IRF, the FLIM analysis software can estimate a response function for the analysis.

Optional: General background can be measured by a 2 min point measurement of water (*see* **Note 3**).

2. The power of the excitation light at the objective should be kept as low as possible to avoid bleaching artifacts during data acquisition. A recommended starting value is 1–2 μW measured at the objective's front lens of a 485 nm laser to excite EGFP in your sample (*see* **Note 4**). However, the excitation power depends on the number of photons that can be detected in an appropriate time. It should ideally be kept constant during a complete set of experiments. At least changing the excitation power should be recorded (*see* **Note 5**).

3.3 FRET-APB Measurements

For FRET-APB measurements, plants are mounted and examined at the microscope (*see* Fig. 5). It is important not to induce acquisition bleaching during the measurement. Therefore, care should be taken to adjust the laser power to the minimum needed. For FRET-APB measurements, it is not necessary to have high-resolution images as the aim is to take the measurements as quickly as possible.

1. The size of the image, pixel time, and laser power should ideally be kept exactly the same for all sets of measurements. In case of higher or lower expressing cells, the detector gain can be adjusted.
2. Because of the spectral overlap of the FPs used as FRET pairs, "cross talk" from one channel to the other has to be avoided. This is very important for FRET-APB as both the donor

(e.g., GFP) channel and the acceptor (e.g., mCherry) channel are imaged simultaneously. Here, a line-wise switching mode of channels is fast and prevents cross-talk problems.

3. In total, 20 frames (256 × 256 pixel) at a pixel time of around 1–3 μs/pixel (high to maximum scan speed) are measured with no line averaging. The whole measurement of 20 frames including the bleaching should be finished within 10–20 s.
4. After the fifth frame, an area is bleached at 100% laser power of the acceptor excitation wavelength, for 50–100 iterations (*see* **Note 6**). The region of interest (e.g. 200 × 50 pixel) can be adjusted, but should be kept the same for the whole set of experiments, as changing it affects the acquisition time.
5. After the bleaching, 15 more frames are recorded. The intensity values of the acquired 20 frames in the two channels can be exported and analyzed according to the formulas (Eq. 2) in the introduction (e.g., in Excel).

3.4 FRET-APB Data Quality Criteria

The most robust and unbiased data can be achieved by two main data quality criteria:

1. The intensities in the first five channels are not supposed to change more than 10% (otherwise acquisition settings have to be adjusted, e.g., to avoid acquisition bleaching artifacts).
2. For calculations of the FRET efficiency E, the mean values of the five measurements before and five measurements after the bleaching should be used. Using only the data point just before and after the bleaching could result in false positive results due to fluorescence recovery after acquisition bleaching.

3.5 FRET-FLM Measurements

For FRET-FLIM measurements, plant tissue is mounted and examined at the microscope as described (*see* Subheading 3.1 and Fig. 5).

The goal of a FRET-FLIM experiment is to measure the decrease of the fluorescence lifetime of the FRET donor fluorophore in the presence of the acceptor fluorophore in comparison to the no-FRET case.

1. The first data set to be acquired should be the "donor-only" sample to get an idea of the initial fluorescence lifetime of the donor fluorophore in case of no FRET in the individual subcellular environment, e.g., the plasma membrane, of the fluorescent fusion protein (*see* **Note 7**). Each data set to be analyzed should consist of at least ten individual measurements; these measurements must be reproducible at several different measuring days.
2. FLIM data sets are acquired by repetitively scanning an image to accumulate a suitable number of photon counts per pixel. The recommended image size is 256 × 256 pixels to restrict the resulting file size for the pixel-wise fluorescence lifetime analysis.

3. The absolute minimum number of photon counts in one pixel should be 100. It is better to have on average 300–400 photon counts per pixel. To apply multicomponent fitting models to the data, the minimum number of counts should be at least 1000 [7]. One measurement can either be stopped after reaching a defined number of photon counts in one pixel or after a certain number of accumulated frames, e.g., 40–60 frames. With a pixel dwell time of around 10 μs, one measurement takes about 1 min. It is better to scan more frames with short pixel dwell times than to scan fewer frames with longer dwell times. This helps to avoid irreversible bleaching of fluorophores that could influence the fluorescence lifetime. If it is necessary to reduce the measurement time of one experiment, e.g., if you want to achieve a better time resolution, it might be practicable to slightly increase the laser power to collect more photon counts in a shorter time. In control experiments it has to be tested if the fluorescence lifetime is not affected by increasing the laser power. After completing the "donor-only" data set, measurements of the FRET sample and further controls are carried out. In FRET-FLIM experiments the most important readout is the fluorescence lifetime of the donor fluorophore; however, it can be helpful to record the fluorescence intensity of the acceptor fluorophore in parallel. Especially in a transient expression system, the amount of expressed proteins can vary from cell to cell and therefore can be used as an internal control. Therefore, we strongly recommend to record in addition to the FLIM a data file where the acceptor fluorophore is excited directly, e.g., in case of FRET between EGFP and mCherry, and a file of ten frames is recorded using 561 nm (continuous wave laser) to excite the mCherry FP.
4. Data analyses of FRET-FLIM measurements are carried out using manufacturer's analysis software tools coming with the FLIM system or independent software tools (*see* **Note 8**). For a first impression, it is sufficient to fit the recorded decay curves to a mono-exponential decay model. FRET is occurring if a reduction in the fluorescence lifetime of the donor in the range of typically 300 ps in the presence of acceptor molecules can be observed (*see* **Note 9**). To quantify the fraction of molecules undergoing FRET in the sample, one has to apply at least a bi-exponential model to describe the decay curve that in addition considers the contribution of background fluorescence and scattered light [13].

3.6 FRET-FLIM Data Quality Criteria

FRET-FLIM measurements in plants generally suffer from a substantial contribution of background fluorescence as they contain a high amount of autofluorescent compounds. Therefore, an exact characterization of the individual autofluorescence in the plant tissue under investigation should be carried out before starting the FRET experiments. For example, chloroplasts have a very high autofluorescence over a broad spectrum with a very short fluorescence lifetime (<1 ns). Thus one should avoid measuring FRET-FLIM in close vicinity to chloroplasts. Furthermore, a high number of photon counts per pixel minimizes the relative contribution of autofluorescence to the decay histogram and helps to increase overall data quality.

3.7 Potential Pitfalls and Points of Consideration

In this section we will focus on, in our experience, important measures to obtain high-quality data. It is very important to avoid any photodamage and/or acquisition bleaching, as this can influence results both in FRET-APB and FRET-FLIM experiments. Always try to use the lowest laser excitation power possible. The general expression level of your plants is of importance, too. Enough donor- and acceptor-labeled proteins have to be expressed; however, strong overexpression of FP-labeled proteins has to be avoided. In our experience, the transient *N. benthamiana* system using an estradiol expression system works very well, and the time of induction can be adjusted to the FP-labeled protein of choice to avoid too low expression or overexpression artifacts [12].

4 Notes

1. Erythrosin B has an extremely short lifetime and can therefore be used to record the instrument response function (IRF) of a FLIM setup [14]. The erythrosin B solution is prepared as follows: Erythrosin B powder is dissolved in distilled water. The solution should be strongly colored. NaI or KI is added until saturation to further quench erythrosin B. The erythrosin B solution is stable for at least 1 month.
2. Each individual FLIM microscope setup creates a characteristic broadening of the exciting laser pulse. This is the so-called instrument response function (IRF) and represents the timing precision of the complete TCSPC system. The IRF is mostly depending on the performance of the detectors and the excitation source. The IRF of your individual system is required for the deconvolution of the decay histogram in the lifetime fitting procedure.
3. This measurement can be used to monitor the dark counts of your detectors as an indicator of their performance. Increase in dark counts can indicate inefficient cooling and aging of detectors. The suppliers of your system distribute typical dark count rates for your type of detectors.
4. Please note that the power density (W/m^2) will increase with zooming into your sample, and unwanted bleaching can occur!

5. Results of calibration measurements and the following measurements should be journalized in a document file in addition to individual data set recording in the FLIM software. This helps to recapitulate all measurement parameters like changing laser excitation power or number of acquired frames and quality of single measurements in the later data analyses.
6. The duration of bleaching has to be tested empirically, as it depends on the laser power of a given system; the goal is to bleach acceptor fluorescence to almost background level as fast as possible.
7. The individual environment of a fluorescent fusion protein can drastically change the fluorescence lifetime of its fluorescent protein. Therefore, it is necessary to carefully investigate the behavior of the fusion protein before starting FRET measurements.
8. Available FLIM analyses software: SymphoTime by PicoQuant (www.picoquant.com), SPCImage by Becker & Hickl (www.becker-hickl.de), FLIMfit (www.openmicroscopy.org), and AnI (www.mpc.hhu.de).
9. The reduction of fluorescence lifetime in case of FRET that can be observed in biological samples can be even lower than 300 ps. Here, autofluorescence, orientation of the fluorophores, and other (maybe unknown) interacting proteins may influence the occurrence of FRET. It is advisable to measure as many samples and combinations as possible to acquire statistically significant results.

Acknowledgments

We acknowledge the Deutsche Forschungsgesellschaft (DFG) for financial support to Y.S. by grant STA12/12 1-1 and S.W.P. by grant WE53/43 1-1. We are thankful to Steffen Köhler for taking photographs of *N. benthamiana* leaves and Rüdiger Simon for critical reading of the manuscript.

References

1. Förster T (1948) Zwischenmolekulare Energiewanderung und Fluoreszenz. Ann Phys 6(2):55–75
2. Albertazzi L, Arosio D, Marchetti L et al (2009) Quantitative FRET analysis with the EGFP-mCherry fluorescent protein pair. Photochem Photobiol 85(1):287–297
3. Lakowicz JR (2006) Principles of fluorescence spectroscopy, 3rd edn. Springer Science+Business Media LLC, Boston, MA
4. Karpova TS, Baumann CT, He L et al (2003) Fluorescence resonance energy transfer from cyan to yellow fluorescent protein detected by acceptor photobleaching using confocal microscopy and a single laser. J Microscopy 209:56–70
5. Widengren J, Terry B, Rigler R (1999) Protonation kinetics of GFP and FITC investigated by FCS — aspects of the use of fluorescent indicators for measuring pH. Chem Phys 249(2–3):259–271
6. Biskup C, Zimmer T, Kelbauskas L et al (2007) Multi-dimensional fluorescence lifetime and FRET measurements. Microsc Res Tech 70(5): 442–451

7. Stahl Y, Grabowski S, Bleckmann A et al (2013) Moderation of Arabidopsis root stemness by CLAVATA1 and ARABIDOPSIS CRINKLY4 receptor kinase complexes. Curr Biol 23(5):362–371
8. Bücherl CA, van Esse GW, Kruis A et al (2013) Visualization of BRI1 and BAK1 (SERK3) membrane receptor heterooligomers during. Plant Physiol 162(4):1911–1925
9. Somssich M, Ma Q, Weidtkamp-Peters S et al (2015) Real-time dynamics of peptide ligand-dependent receptor complex formation in planta. Sci Signal 8(388):ra76
10. Shaner NC, Steinbach PA, Tsien RY (2005) A guide to choosing fluorescent proteins. Nat Methods 2(12):905–909
11. Piston DW, Kremers G (2007) Fluorescent protein FRET: the good, the bad and the ugly. Trends Biochem Sci 32(9):407–414
12. Bleckmann A, Weidtkamp-Peters S, Seidel CAM et al (2010) Stem cell signaling in Arabidopsis requires CRN to localize CLV2 to the plasma membrane. Plant Physiol 152(1): 166–176
13. Weidtkamp-Peters S, Felekyan S, Bleckmann A et al (2009) Multiparameter fluorescence image spectroscopy to study molecular interactions. Photochem Photobiol Sci 8(4):470–480
14. Boens N, Qin W, Basarić N et al (2007) Fluorescence lifetime standards for time and frequency domain fluorescence spectroscopy. Anal Chem 79(5):2137–2149

Chapter 17

Detection and Analyses of Endocytosis of Plant Receptor Kinases

Jenna Loiseau and Silke Robatzek

Abstract

Genetic dissection has led to a sophisticated understanding of receptor kinases in plant development and responses to abiotic and biotic stresses. Fluorescence confocal microscopy is essential to identify the (sub) cellular locations of resting and signaling receptor kinases that trigger molecular events in plant cells upon ligand perception. In this regard, the internalization of plasma membrane-localized FLAGELLIN SENSING 2 (FLS2) into endosomes induced by its ligand flg22, a peptide derived from bacterial flagellin, is a model system for studying activation status-dependent and endosomal receptor kinase trafficking routes and can be used in screens to identify pathogen effectors that target these trafficking routes for virulence promotion. In this chapter we describe approaches of visualizing fluorescently tagged FLS2, including protocols for flg22-induced endocytosis, instrument parameters, and image analysis. These approaches can be easily adapted for other receptor kinases, using the fast transient expression system in *Nicotiana benthamiana* for microscopic inspection.

Key words Confocal microscopy, Live-cell imaging, FLS2, Flagellin, flg22, Endocytosis, *Nicotiana benthamiana*

1 Introduction

Innate recognition of potentially pathogenic microbes is mediated by pattern recognition receptors (PRRs) localized at the plasma membrane and activates the plant immune system [1]. Prominent members of PRRs belong to the receptor kinase or receptor-like protein families. FLAGELLIN SENSING 2 (FLS2) is a well-characterized receptor kinase that detects bacterial flagellin through its conserved flg22 epitope, mediating antibacterial immunity [2]. Induced by flg22, FLS2 forms a signaling-competent receptor complex with the co-receptor kinase BRASSINOSTEROID INSENSITIVE 1 (BRI1)-ASSOCIATED KINASE 1 (BAK1) [3] triggering cellular events that, among others, result in the translocation of plasma membrane-localized FLS2 into endosomes [4]. Ligand-dependent endocytosis is conserved across PRRs. In addition

Reidunn Birgitta Aalen (ed.), *Plant Receptor Kinases: Methods and Protocols*, Methods in Molecular Biology, vol. 1621,
DOI 10.1007/978-1-4939-7063-6_17, © Springer Science+Business Media LLC 2017

to FLS2, these include the receptor kinases EF-TU RECEPTOR (EFR) recognizing bacterial EF-TU, PEP RECEPTOR 1 (PEPR1) that perceives damage-associated endogenous peptides [5], as well as complexes of the SUPPRESSOR OF BIR1–1 (SOBIR1) receptor kinase that associates with receptor-like proteins such as the tomato Cf receptors to mediate antifungal immunity [6, 7]. Utilizing functional fusions of these receptor kinases to fluorescent proteins, their (sub)cellular localization patterns can be visualized by confocal microscopy and resolved in a ligand-dependent and dynamic manner [8].

Dissecting the pathways that remove PRRs from the plasma membrane is critical for our understanding of how plants respond to microbial perception. For example, FLS2-positive endosomes can be rerouted to pathogen infection sites [9]. To advance the time-intensive limitations that typically occur from generating stable transgenic plants, *Agrobacterium tumefaciens*-mediated transient transformation of *Nicotiana benthamiana* leaves is a powerful system for simple and fast genetically encoded expression. This approach is routinely used, e.g., to monitor bimolecular fluorescence complementation [10] and assess the localization and function of virulence proteins secreted by pathogens to suppress plant defenses [9, 11–13]. We have adapted heterologous expression in *N. benthamiana* leaves combined with confocal microscopy to dissect the endocytic routes of PRR kinases tagged with fluorescent proteins. Co-expression with different fluorescently tagged markers of defined subcellular compartments, e.g., plasma membrane (PM), Golgi, *trans*-Golgi network (TGN), and multivesicular bodies (MVBs), allows the probing of pattern recognition receptor localization along distinct trafficking routes and during infection [10]. In addition, genetic interference by overexpressing dominant-negative variants of trafficking regulators and virus-induced and RNAi-mediated gene silencing allows the identification of molecular determinants involved in ligand-induced endocytosis of pattern recognition receptor kinases [5, 10, 12]. Likewise, co-expression of pathogen virulence proteins can be used to interfere with receptor kinase localization [12], thus providing insights into the mechanisms by which pathogens reprogram plant cellular responses. For example, FLS2 endocytosis can be inhibited by co-expression of *Phytophthora infestans* AVR3a, a virulence protein interacting with *N. benthamiana* dynamin-related protein 2 (DRP2) [12].

In this chapter, we provide instructions for transient expression of FLS2-GFP in *N. benthamiana* leaves and how to monitor FLS2-GFP localization by confocal microscopy. We describe how to ensure correct FLS2-GFP expression and imaging of FLS2-GFP fluorescent signals and provide tools for quantification of endosomes from images. Although we focus on FLS2-GFP in this chapter, this method can be broadly applied for imaging other fluorescent-tagged receptor kinases in *N. benthamiana* [10] and combined with co-expression of subcellular markers, pathogen virulence proteins, and genetic interference to functionally dissect their dynamic (sub)cellular localization.

2 Materials

2.1 Samples

1. Four-week-old *Nicotiana benthamiana* plants (*see* Fig. 1) grown on soil under 16 h light at 22 °C/80% humidity.
2. *pFLS2:FLS2-3xmyc-GFP* cloned in pCAMBIA2300 [14] (*see* Table 1).
3. *Agrobacterium tumefaciens* strain GV3101 carrying pMP90.

2.2 Experiment/ Treatments

1. LB medium: Tryptone 10 g/L, yeast extract 5 g/L, sodium chloride 10 g/L.
2. Antibiotics: rifampicin, prepare stock solution at 50 mg/L in DMSO; gentamicin, stock solution at 30 mg/L in water; and kanamycin, stock solution at 50 mg/L in water; all are kept at −20 °C.

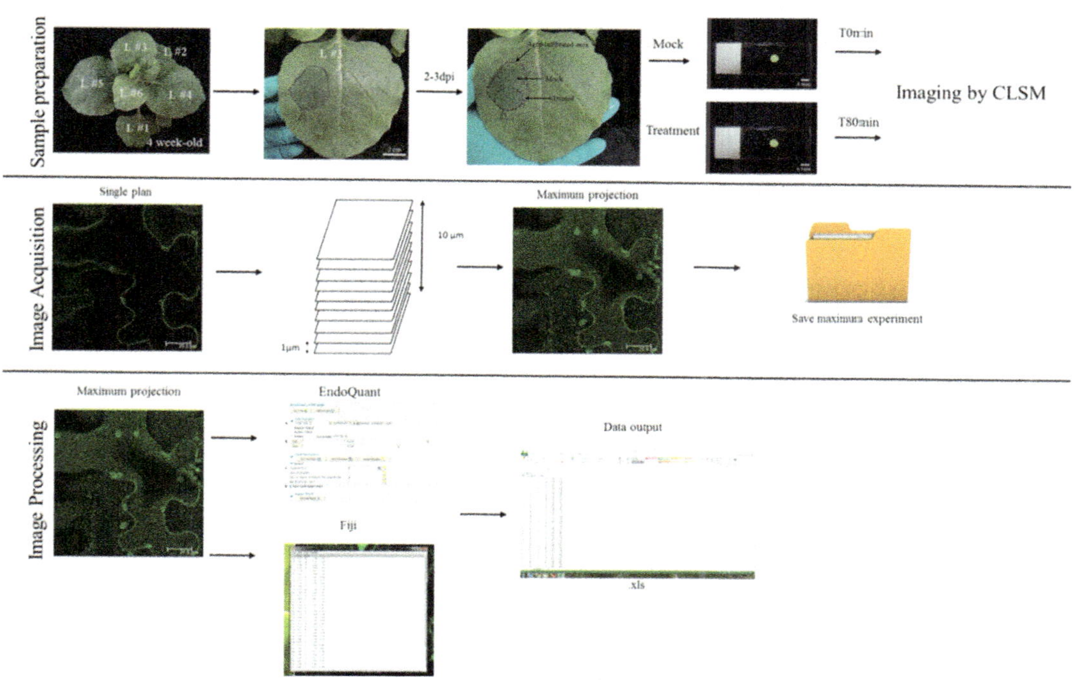

Fig. 1 General workflow. For *sample preparation*, leaf disks are mounted on microscopy slides and cover glass in water and 100 µM flg22 for mock and treatment, respectively. Sample treatments are performed by infiltration in the epidermal tissue before mounting process. The *image acquisition* is performed with a confocal laser scanning microscope (CLSM). The generated files are analyzed with EndoQuant or with Fiji. The *data output* is generated in table format

Table 1
Receptor kinase localization in *N. benthamiana*

Marker	Organelle[a]	Heterologous expression	Excitation-emission	*N. benthamiana* orthologues[c]	Reference
ACA8-mCherry	PM[a]	2–3 dpi[b]	561 nm–580/620 nm	Niben101Scf04852g01008.1	[10]
mCherry-MEMB12	Golgi	2–3 dpi	561 nm–580/620 nm	No hits	[10]
mRFP-SYP61	TGN	2–3 dpi	561 nm–580/620 nm	Niben101Scf02944g02004.1	[15]
VHA-a1-RFP	TGN	2–3 dpi	561 nm–580/620 nm	Niben101Scf11756g01025.1	[22]
RFP-ARA7	Endosomes	3 dpi	561 nm–580/620 nm	Niben101Scf02976g01015.1Niben101Scf00271g01020.1	[22]
ARA6-RFP	LE	2–3 dpi	561 nm–580/620 nm	Niben101Scf29276g00003.1Niben101Scf00648g00003.1	[22]
RFP-VPS37-1	MVB	3 dpi	561 nm–580/620 nm	No hits	[22]
YFP-RabG3c	Tonoplast, vacuole	3 dpi	488 nm–495/550 nm	Niben101Scf05709g00001.1Niben101Scf01374g03034.1 Niben101Scf07008g01002.1	[9]
RFP-RabG3c	Tonoplast, vacuole	3 dpi	561 nm–580/620 nm		

[a] *PM* plasma membrane, *TGN* trans-Golgi network, *LE* late endosomes, *MVB* multivesicular body
[b] *dpi* day post infiltration
[c] Putative *N. benthamiana* orthologues obtained by sequence similarity (BLAST) against *N. benthamiana* v1.0.1 predicted cDNA

3. 3′,5′-Dimethoxy-4′-hydroxyacetophenone (acetosyringone): stock solution of 250 mM in DMSO, use 100 μM final concentration in water.
4. flg22 peptide: QRLSTGSRINSAKDDAAGLQIA (custom produced), MW = 2272.5 g/L.

2.3 Confocal Microscopy

Sample Mounting

Samples are mounted in water or 100 μM flg22 solution for mock or for treated conditions, respectively. Leaf disks are excised from *A. tumefaciens*-infiltrated leaves with a cork borer No. 3 (ø 6.5 mm) and mounted between a cover glass (22 × 50 mm, R&L Slaughter Ltd., 631 0137) and glass microscope slide (76 × 26 mm SKAN LTD).

Image Acquisition

1. Confocal laser scanning microscope.
2. Lasers: Argon ion, DPSS 561.
3. Detectors coupled with cameras.
4. Scan parameters: Acquisition, *xyz* for single plane and for *z* sectioning and *xyt* for time-lapse movies. Format: 512 × 512 pixel, scan speed 400 Hz (400 lines/s); pinhole size as default at 1 Airy Unit (AU). Averaging: line 1×, frame 1× (*see* **Note 1**).
5. Objectives: Start with a low magnification to find the sample (e.g., 10×), and then switch to higher magnification to detect spots (63×).
6. Zoom factor: 2× zoom, image size 122 μm × 122 μm.

Image Processing

1. EndoQuant
   EndoQuant is a modification of EndomembraneQuantifier [4], runs on the PerkinElmer Acapella image analysis software package, and can be used for spot detection and quantification for standard confocal images [10].
2. FIJI
   FIJI ("FIJI Is Just ImageJ") is an open-source platform for biological image analysis that comes preloaded with an extended set of commonly used plug-ins:
   FIJI homepage: http://pacific.mpi-cbg.de/wiki/index.php/Fiji
   ImageJ homepage: http://rsb.info.nih.gov/ij/

3 Methods

General workflow, *see* Fig. 1.

Sample preparation and treatments, image acquisition, and image processing.

3.1 Samples

1. For transient expression of FLS2 fused to the green fluorescent protein (GFP) in *N. benthamiana*, the pCAMBIA2300 plasmid carrying pFLS2:FLS2-3xmyc-GFP is introduced into *Agrobacterium tumefaciens* GV3101 by transformation. Leaves of *N. benthamiana* will be used for transient transformation. Imaging is performed on leaf disks.
2. Inoculate 10 ml LB medium supplemented with antibiotics (rifampicin 50 mg/L, gentamicin 30 mg/L, kanamycin 50 mg/L) with *A. tumefaciens* GV3101 carrying *pFLS2:FLS2-3xmyc-GFP* and incubate for 16 h shaking at 28 °C.
3. Centrifuge the bacterial culture for 10 min at 5311 RCF. Remove supernatant and resuspend the pellet in 5 ml dH_2O (*see* **Note 2**).
4. Take 100 μl of bacterial suspension and dilute 1/10 in water to measure OD_{600}.
5. Prepare 5–10 ml of bacterial suspension at final OD_{600} = 0.1–0.2 in dH_2O and add 100 μM 3′,5′-dimethoxy-4′-hydroxyacetophenone (acetosyringone). Incubate for 1 h at RT in the dark (*see* **Note 3**).
6. Turn leaf #3 or #4 of a 4-week-old *N. benthamiana* plant (*see* Fig. 1) to face its abaxial side upward and pinch the leaf carefully with a needle.
7. Where the leaf has been pinched, carefully inject the bacterial suspension ("infiltration") using a needleless syringe, and fill about 0.5 ml and infiltrated leaf (*see* **Notes 4** and **5**).
8. Mark the inoculated area with a soft permanent marker on the leaf apical side, and incubate plants in growth chamber for 1–3 days (*see* **Note 6**).
9. To observe localization of non-liganded and liganded, activated FLS2-GFP in *N. benthamiana*, perform co-localization studies (*see* Table 2). Organelle markers are heterologously co-expressed by *A. tumefaciens*-mediated transient transformation as described above. To express several constructs (up to four), premix *Agrobacterium* suspensions to infiltrate each construct at OD_{600} = 0.1–0.4 (*see* **Note 7**).

3.2 Treatment

1. Non-liganded FLS2-GFP resides predominantly in the plasma membrane [15] and co-localizes with the plasma membrane marker ACA8-mCherry [8].
2. Liganded, activated FLS2-GFP is detected at endosomes around 80–90 min after flg22 treatment and observed as mobile spots. Endosomal FLS2-GFP co-localizes with mRFP-SYP61 at 90–120 min after flg22 treatment and localizes to RFP-ARA7/RabF2b-positive late endocytic compartments from 30 up to 200 min after flg22 treatment [15]. Additional markers can be used to trace FLS2-GFP endocytic trafficking (*see* **Note 8**, Table 2) [16–18].

Table 2
Markers used for co-localization in heterologously expressing *N. benthamiana* leaves

Receptor kinase	Organelle	Heterologous expression	Excitation-emission	Reference
FLS2-GFP	PM[a], endosomes	2–3 dpi[b]	488 nm–495/550 nm	[15]
EFR-GFP	PM, endosomes	2–3 dpi	488 nm–495/550 nm	[8]
BRI1-GFP	PM, endosomes	2–3 dpi	488 nm–495/550 nm	[8]
PEPR1-YFP	PM, endosomes	2–3 dpi	514 nm–520/560 nm	[8]
SOBIR1-GFP	PM, endosomes	2–3 dpi	514 nm–520/560 nm	[7, 10]

[a]*PM* plasma membrane
[b]*dpi* days post infiltration

Table 3
Genetic interference of membrane trafficking in *N. benthamiana* leaves

Construct[a]	Target	Effect on FLS2 localization	FLS2 localization after flg22 treatment	Reference
DN-RABA6ab^{N126I}	RABA6a	Delayed maturation from TGN to LE	TGN	[15]
DN-RABA4c^{N128I}	RABA4c	Inhibition of FLS2 transport to the TGN	Endosomes	
TRV::*Nb*SERK3a/b	*Nb*SERK3a/b	Inhibition of FLS2 internalization	PM	[8, 10]
hp*Nb*CHC	6 *Nb*CHCs	Inhibition of FLS2 internalization	PM	[8]
hp*Nb*DRP2	*Nb*05397 *Nb*31648	Reduced number of FLS2-GFP punctae	PM, punctae	[12]
AVR3a	DRP2	Reduced number of FLS2-GFP punctae	PM, punctae	[12]

[a]*TRV* Tobacco Rattle Virus, *hp.* hairpin
[b]*PM* plasma membrane, *LE* late endosomes, *TGN trans*-Golgi network

3. To validate that the observed mobile spots are bona fide endosomes and not, e.g., secretory vesicles, co-localization experiments using the Golgi marker mCherry-MEMB12 should be negative [10, 18].
4. FLS2-GFP endocytic trafficking can be genetically dissected using transient-silencing approaches, overexpression of dominant-negative trafficking regulators, and co-expression of pathogen virulence proteins (*see* Table 3; [10, 12, 15]). For example, the *P. infestans* effector AVR3a has been identified to target *N. benthamiana* DRP2 [12]. In agreement with

ligand-induced endocytosis depending on DRP2b in *Arabidopsis* [19], co-expression of AVR3a impaired FLS2 endocytosis [12]. Consistently, using a hairpin-based RNA-mediated-silencing approach, FLS2 endocytosis was affected upon knockdown of *N. benthamiana* DRP2 expression [12]. This demonstrates that heterologous expression in *N. benthamaina* leaves is a suitable system to dissect endocytic trafficking of receptor kinases.

3.3 Experiment/Treatment

1. Excise leaf disks with a cork borer No. 3 (ø 6.5 mm) from the inoculated area. Drop 70 μl of dH_2O in the middle of a cover glass. Using tweezers to grasp the disk edge, place a single-leaf disk on the liquid drop with the abaxial surface facing down. Add two drops of 60 μl water at each side of the disk. Cover with microscope slide. Carefully invert the mounted leaf disk. Abaxial leaf side is now facing up. Fill the remaining space between slide and cover glass with water. Image samples to observe FLS2-GFP at the plasma membrane.
2. For flg22 treatments, pinch the abaxial leaf side of the inoculated area close to where it has been done before for bacterial inoculation. Using a 1 ml needleless syringe, infiltrate ca. 0.1 ml of 100 μM flg22 solution filling an area about ø 3.5 cm. Mark the infiltrated area on the apical leaf side. Incubate at RT for ca. 70–90 min. Then, excise leaf disks with a cork borer and mount the disks as described above in 100 μM flg22 solution instead of water, to ensure continued treatment. Image samples between 80 and 200 min of flg22 treatments to observe FLS2-GFP-positive endosomes.

3.4 Confocal Microscopy

Image Acquisition

1. Excitation of the samples is performed with the 488 nm argon laser for GFP, and emission is collected between 495 and 550 nm. For co-localization studies and use of other fluorescent proteins, *see* Table 2.
2. A water immersion 63x/NA1.20 objective is used for subcellular imaging of FLS2-GFP on the abaxial side of the sampled leaf disk. Depending on the microscope lasers, detectors, camera, and/or resolution, unspecific signals may be recorded. Therefore, when setting up the system, it is critical to image *N. benthamiana* leaves that were inoculated with *A. tumefaciens* not carrying a vector with FLS2-GFP, preferably an empty vector. Following the microscopy instructions, set up the imaging parameters such that epidermal cells are clearly detected and in focus when viewing the bright field channel and such that autofluorescence of chloroplasts is detected in the autofluorescence control channel (e.g., 700–800 nm), but

no or negligible signal should be observed in the GFP channel. Using these settings, start imaging disks from FLS2-GFP-expressing *N. benthamiana* leaves. It might be necessary to adapt the image acquisition settings in order to optimize detection in the GFP channel, but when imaging leaf disks that were not transformed with FLS-GFP, no or negligible signal should be observed using the same settings (*see* **Notes 9** and **10**).

3. Epidermal puzzle-shaped cells are visualized. In the acquisition mode, select *xyz* to allow acquisition of z-stacks. Choose scan parameters: Tick the pinhole box to control the image contrast, format 512 × 512 px, 400 Hz speed, line averaging 1×, and frame averaging 1× (*see* **Note 1**). Unidirectional scanning is best for image quality, avoiding artifacts that originate from interlacing after bidirectional scanning. Select cells with good GFP signal at the plasma membrane (*see* Fig. 1; *see* **Notes 11** and **12**). Perform a 2× zoom (area size, 122 μm × 122 μm). Take a z-stack from to the top of the cell downward (*see* **Note 11**). Take ~10 z-sections that are 1 μm separated (*see* **Note 11**). FLS2-GFP-positive endosomes can be observed as mobile spots (*see* **Note 7**) and co-localize with endosomal markers (*see* Table 2). To ensure that the signal observed is truly GFP, perform a lambda wavelength scan according to microscopy instructions.
4. Save your experiments (format depends on microscope manufacturer).

Image Processing

Confocal images can be processed automatically using EndoQuant or manually with FIJI.

1. EndoQuant

EndoQuant is a modification of EndomembraneQuantifier [4], runs within the PerkinElmer Acapella image analysis software package, and can be used for spot detection and quantification for standard confocal images [10].

(a) Generate maximum projections of confocal z-stacks and save as PNG, TIFF, JPG, or BMP files.

(b) Place all resulting images to be analyzed in a single folder.

In order to run EndoQuant, you need the PerkinElmer Acapella image analysis program.

(a) Start Acapella.

(b) Open EndoQuant, for example, by dragging the .script file onto the open Acapella screen.

(c) Set "Data Selection" to "Single Step."

(d) Set "Path" and "Image Directory" to the location of the folder with images to be analyzed.

(e) Set "Select Input Image Format" to match the format of images to be analyzed.

(f) It is recommended to switch on "Remove Objects Attached to The Image Border" and switch off "Background very noisy?," but these can be adjusted based on preference.

(g) Click "Run Script."

EndoQuant generates comma-separated value .csv files, which can be opened using MS Excel or other spreadsheet software. For each analyzed confocal micrograph, EndoQuant also generates images that show which spots were detected, which size they were classified as (red, big; yellow, medium; green, small), and how they were numbered. These images can be used as a visual evaluation of the quality of spot detection.

(a) Go to the automatically generated "results" folder in the image folder that was analyzed.

(b) Open "overall results.csv" and separate values in column A based on symbol "#."

(c) In the resulting spreadsheet, the column "Endosome_No" can be found, and this data is used as a value for number of endosomes in each individual confocal micrograph.

(d) The detected spots are further separated into three size classes for more detailed information on size distribution.

(e) "Endosome_No" values can be used for further statistical data analysis.

2. FIJI

FIJI is an open-source platform for biological image analysis and can read most confocal microscope data formats using the built-in BioFormats plug-in. Maximum projections are opened and processed with FIJI.

(a) Launch the FIJI software and open the saved experiment and process as a maximum projection.

(b) Open the maximum projection in FIJI to analyze and quantify spots.

(c) Within the FIJI menu, right-click on the multipoint tool. A window will appear that allows setting the parameters of the selection (type, circle; color, magenta; and size, large; tick the box label points).

(d) Select and click on spots/endosomes with pointer. Numbers appear as you click (selection).

(e) Upon completion of manual spot detection, extract the information. In the Analyze tab click on "Measure" or press Ctrl + M: a table will appear in a new window. Number of spots and the corresponding coordinates will be listed in the table. In the "File" tab of the results window, click on save as all file *.*. The table is saved and can be later opened with Excel or other spreadsheet software.

(f) To save the circled spots (selection) on the maximum projection as an image, the selection created must be added to the maximum projection as an overlay (Image overlay, add selection, or CTRL + B), and the overlay must be flattened (Image, Overlay, Flatten, or Ctrl + Shift + F).

4 Notes

1. To enhance FLS2-GFP signals and reduce noise, background, or blurry signals, scan parameters can be adjusted (e.g., change to format 1024 × 1024 px, 200 Hz speed, line average 3×). However, enhanced laser power or longer local exposure can damage leaf cells, which in turn induces the accumulation of autofluorescing compounds recorded as false-positive signal, introduce stresses that may cause aberrant receptor localization, and bleach the GFP signal.
2. Transient expression can trigger cell death symptoms. To avoid triggering cell death, remaining LB medium and antibiotics should be removed by a second wash.
3. To enhance transient transformation efficiency, OD_{600} or 3′,5′-dimethoxy-4′-hydroxyacetophenone (acetosyringone) concentration can be increased.
4. To improve bacterial inoculation, plants can be watered or humidified 1–2 h prior the infiltration.
5. To enhance transient expression of p35S-driven constructs, the plant viral protein p19-silencing suppressor [21] can be co-inoculated at $OD_{600} = 0.2$.
6. Expression levels and accumulation of the full-length fluorescent-tagged FLS2-GFP (other receptor kinase fluorescent fusions and organelle marker fluorescent fusions; *see* Tables 1 and 2) should be validated by immunoblot analysis for different OD_{600} (0.1–0.5) and time points after inoculation (e.g., 1–3 dpi) prior to confocal microscopy.
7. The extent to which constructs are expressed in transiently transformed *N. benthamiana* leaves varies between individual cells, leaves, and plants. Thus, when comparing mock and treated conditions, use samples from the same leaf. Such within-leaf comparisons are critical when quantifying

FLS2-GFP endosomes from genetic interference experiments (*see* Fig. 1).

8. Co-localization studies with organelle markers are performed to observe FLS2-GFP transit through the late endosomal pathway after activation and to dissect its trafficking (*see* Table 2).
9. Autofluorescence of chloroplasts is collected between 700 and 800 nm. This is important to record in the red channel as it can also be detected in the GFP channel and misinterpreted as signal from receptor kinase GFP fusions. Likewise, chlorophyll autofluorescence might be misinterpreted as RFP/mCherry signal when using those fluorophores to image organelle markers.
10. Save the settings used to image the control material (laser power and intensity, pinhole, gain, zoom, numbers of z-sections, slide size), and use the same parameters to acquire the signal of the treated experiment to prevent variation due to confocal settings.
11. Since epidermal cells of *N. benthamiana* mature leaves are highly vacuolated, FLS2-GFP-positive endosomes are best observed close to the cell periphery. Therefore, it is recommended to acquire z-stacks of ca. 10 μm depth and to include the top of cells where the orientation of the periphery is horizontal. Endosomes are visible as mobile signal-positive punctae. Acquire time-lapse movies to capture vesicle mobility. For this, choose a single plane where spots are detected (e.g., near the top of cells), select *xyt* mode in acquisition parameters, and record a movie for up to 1 min (movies are saved as .avi files).
12. Interpretation of localization or co-localization results must be done cautiously and insure that the tag does not affect the protein function. As an example c-terminally tagged BAK1 is not functional in PTI signaling but functional in BR signaling [20]. Testing functionality of tagged protein function can be performed. For instance, *Nb*SERK3a/b can be silenced and functionally complemented with untagged *At*BAK1 [10].

References

1. Boller T, Felix G (2009) A renaissance of elicitors: perception of microbe-associated molecular patterns and danger signals by pattern-recognition receptors. Annu Rev Plant Biol 60(1):379–406
2. Zipfel C, Robatzek S, Navarro L, Oakeley EJ, Jones JDG, Felix G, Boller T (2004) Bacterial disease resistance in Arabidopsis through flagellin perception. Nature 428(6984):764–767
3. Chinchilla D (2007) A flagellin-induced complex of the receptor FLS2 and BAK1 initiates plant defence. Nature 448:497–500
4. Beck M, Zhou J, Faulkner C, MacLean D, Robatzek S (2012) Spatio-temporal cellular dynamics of the arabidopsis flagellin receptor reveal activation status-dependent endosomal sorting. Plant Cell 24(10):4205–4219
5. Frescatada-Rosa M, Robatzek S, Kuhn H (2015) Should I stay or should I go? Traffic

control for plant pattern recognition receptors. Curr Opin Plant Biol 28:23–29

6. Liebrand TWH, van den Berg GCM, Zhang Z, Smit P, Cordewener JHG, America AHP, Sklenar J, Jones AME, Tameling WIL, Robatzek S, Thomma BPHJ, Joosten MHAJ (2013) Receptor-like kinase SOBIR1/EVR interacts with receptor-like proteins in plant immunity against fungal infection. Proc Natl Acad Sci U S A 110(24):10010–10015
7. Peng K-C, Wang C-W, Wu C-H, Huang C-T, Liou R-F (2015) Tomato SOBIR1/EVR homologs are involved in elicitin perception and plant defense against the oomycete pathogen *Phytophthora parasitica*. Mol Plant-Microbe Interact 28(8):913–926
8. Mbengue M, Bourdais G (2016) Clathrin-dependent endocytosis is required for immunity mediated by pattern recognition receptor kinases. Proc Natl Acad Sci U S A 113(39): 11034–11039
9. TO B, Belhaj K, Dagdas YF, Chaparro-Garcia A, Wu C-H, Cano LM, Kamoun S (2015) Rerouting of plant late endocytic trafficking toward a pathogen interface. Traffic 16(2): 204–226
10. Postma J, Liebrand TWH, Bi G, Evrard A, Bye RR, Mbengue M, Kuhn H, Joosten MHAJ, Robatzek S (2016) Avr4 promotes Cf-4 receptor-like protein association with the BAK1/SERK3 receptor-like kinase to initiate receptor endocytosis and plant immunity. New Phytol 210(2):627–642
11. TO B, Schornack S, Win J, Shindo T, Ilyas M, Oliva R, Cano LM, Jones AME, Huitema E, van der Hoorn RAL, Kamoun S (2011) *Phytophthora infestans* effector AVRblb2 prevents secretion of a plant immune protease at the haustorial interface. Proc Natl Acad Sci U S A 108(51):20832–20837
12. Chaparro-Garcia A, Schwizer S, Sklenar J, Yoshida K, Petre B, Bos JIB, Schornack S, Jones AME, TO B, Kamoun S (2015) *Phytophthora infestans* RXLR-WY effector AVR3a associates with dynamin-related protein 2 required for endocytosis of the plant pattern recognition receptor FLS2. PLoS One 10(9):e0137071
13. Dagdas YF, Belhaj K, Maqbool A, Chaparro-Garcia A, Pandey P, Petre B, Tabassum N, Cruz-Mireles N, Hughes RK, Sklenar J, Win J, Menke F, Findlay K, Banfield MJ, Kamoun S, TO B (2016) An effector of the Irish potato famine pathogen antagonizes a host autophagy cargo receptor. eLife 5:e10856. doi:10.7554/eLife.10856
14. Robatzek S, Chinchilla D, Boller T (2006) Ligand-induced endocytosis of the pattern recognition receptor FLS2 in Arabidopsis. Genes Dev 20(5):537–542
15. Choi SW, Tamaki T, Ebine K, Uemura T, Ueda T, Nakano A (2013) RABA members act in distinct steps of subcellular trafficking of the FLAGELLIN SENSING2 receptor. Plant Cell 25(3):1174–1187
16. Spallek T, Beck M, Ben Khaled S, Salomon S, Bourdais G, Schellmann S, Robatzek S (2013) ESCRT-I mediates FLS2 Endosomal sorting and plant immunity. PLoS Genet 9(12):e1004035. doi:10.1371/journal.pgen.1004035
17. Dettmer J, Hong-Hermesdorf A, Stierhof Y-D, Schumacher K (2006) Vacuolar H+-ATPase activity is required for endocytic and secretory trafficking in Arabidopsis. Plant Cell 18(3): 715–730
18. Geldner N (2009) Cell polarity in plants; a PARspective on PINs. Curr Opin Plant Biol 12(1):42–48
19. Smith JM, Leslie ME, Robinson SJ, Korasick DA, Zhang T, Backues SK, Cornish PV, Koo AJ, Bednarek SY, Heese A (2014) Loss of *Arabidopsis thaliana* Dynamin-related protein 2b reveals separation of innate immune signaling pathways. PLoS Pathog 10(12):e1004578. doi:10.1371/journal.ppat.1004578
20. Ntoukakis V, Schwessinger B, Segonzac C, Zipfel C (2011) Cautionary notes on the use of C-terminal BAK1 fusion proteins for functional studies. Plant Cell 23(11):3871–3878
21. Ding S-W, Voinnet O (2007) Antiviral immunity directed by small RNAs. Cell 130(3): 413–426
22. Lu Y-J, Schornack S, Spallek T, Geldner N, Chory J, Schellmann S, Schumacher K, Kamoun S, Robatzek S (2012) Patterns of plant subcellular responses to successful oomycete infections reveal differences in host cell reprogramming and endocytic trafficking. Cell Microbiol 14(5):682–697

Index

Reidunn Birgitta Aalen (ed.), *Plant Receptor Kinases: Methods and Protocols*, Methods in Molecular Biology, vol. 1621, DOI 10.1007/978-1-4939-7063-6, © Springer Science+Business Media LLC 2017

CPSIA information can be obtained
at www.ICGtesting.com
Printed in the USA
BVOW07*0223130617
486752BV00002B/24/P

9 781493 970629